ExoticsCon Virtual 2020

Building Exotics Excellence

Online
11 September – 3 October 2020

ISBN: 978-1-7138-3038-2

Printed from e-media with permission by:

Curran Associates, Inc.
57 Morehouse Lane
Red Hook, NY 12571

CURRAN ASSOCIATES INC. proceedings.com

Some format issues inherent in the e-media version may also appear in this print version.

Association of Reptilian & Amphibian Veterinarians
5003 SW 41st Blvd
Gainesville FL 32608-4930
USA

Phone: 480-703-4941

info@arav.org

Additional copies of this publication are available from:

Curran Associates, Inc.
57 Morehouse Lane
Red Hook, NY 12571 USA
Phone: 845-758-0400
Fax: 845-758-2633
Email: curran@proceedings.com
Web: www.proceedings.com

EXOTICSCON VIRTUAL

Building Exotics Excellence.

Sept 11 - Oct 3, 2020

2020

AAV Association of Avian Veterinarians

AEMV Association of Exotic Mammal Veterinarians

ARAV Association of Reptile and Amphibian Veterinarians

AAV

Companion Parrot Nutrition—Nutrient Requirements, Common Feeding Practices, and Nutrition-Related Diseases

ExoticsCon Virtual 2020 Proceedings

Laurie Hess, DVM, DABVP (Avian Practice)
Veterinary Center for Birds & Exotics, Bedford Hills, NY, USA; ZuPreem, Premium Nutritional Products, Shawnee, KS, USA

Abstract

This masterclass will review recommended nutrient requirements for psittacine birds and how these requirements were established. The class will also cover what the critical nutrients in companion parrots' diets are and the health implications of their lacking these critical nutrients. Common feeding practices of parrot owners in the U.S. will be discussed, as documented by the surprising findings of two recent large feeding surveys of U.S. parrot owners. Finally, the lecture will cover significant nutrition-related diseases (such as atherosclerosis and gout) associated with these common feeding practices. We will also review the clinical relevance of these practices and what veterinarians and veterinary technicians can do to help bird owners feed their pets more complete nutrition so that companion birds live longer, healthier lives.

Introduction

Despite improvements in feeding practices by pet bird owners during the past two decades, companion parrots still suffer from many nutrition-related diseases that are recognized by avian veterinarians worldwide. The exact nutrient requirements for psittacine birds are not known, and dietary recommendations for these species have been extrapolated from studies done with poultry. To better educate bird owners, it is essential that veterinary professionals are familiar with which nutrients are essential to the health of pet birds, how pet birds' nutritional needs differ from those of their wild counterparts, and what happens when these nutritional needs are not met. With this knowledge, veterinary staff can help educate bird owners about feeding their pets to prevent nutrition-related illnesses.

Feeding Strategies of Psittacine Birds

Psittacine birds are herbivores (consume plants) and are further subdivided based on general food consumption:[1]

- Granivores (budgerigars, cockatiels, hyacinth macaws)
- Frugivores (some macaws, Amazon parrots)
- Nectivores (lorikeets, lories)
- Omnivores (sulphur-crested cockatoos, African gray parrots)
- Mixed (Scarlet macaw is a frugivorous granivore)

In all subdivisions, food consumed may vary based on nutrient availability, age, and sex.

Nutrients

Definition

Dietary components that provide energy and precursors for the synthesis of the body's structural and functional macromolecules

- Macronutrients (lipids, protein, carbohydrates, water)
- Micronutrients (vitamins and minerals)

Essential Nutrients

- Required for optimal health
- Needed for metabolism
- Not synthesized adequately by the body
- Generally known for domestic Galliformes and extrapolated to other species

Nutrient Requirements of Birds

- Few nutrient requirements have been determined scientifically
- The minimal level that optimizes a bird's health is considered the requirement
- Nutrient requirements are determined for three physiologic states:
  - Basal: Those needed to maintain basic life functions
  - Maintenance: Those needed for basal functions plus maintaining body temperature, interacting with other animals, finding food
  - Total: All those needed for life at various stages (growth, reproduction, molting)
- Few studies have been done to establish nutrient requirements of psittacine and passerine species
- Nutrient requirements of companion birds have been extrapolated from those of Galliformes
- Some nutritional information has been extrapolated from studies of budgerigars and cockatiels
- These extrapolations do not account for several variables:
  - Differences in efficiency of nutrient absorption between species
  - Variable nutrient requirements for different species
  - Variable requirements for different life stages and activities:
    - Breeding/reproduction
    - Exercise/flight/migration
    - Growth
    - Molting
    - Aging

The Importance of Vitamins

Vitamin A

- Ingested as beta carotene and other carotenoids
- Degraded enzymatically in intestinal epithelium to retinal which is converted to retinol and retinoic acid
- Absorbed as retinol and retinol esters from GI tract
- Stored in liver
- Commonly deficient in diets of both captive and wild birds
- Exact requirement in psittacine species is unknown
- Feeding trials in female cockatiels showed that birds thrived at 2000–10,000 IU/kg diet
- Requirements in other species less dependent on grain in diet may be higher

- Deficiency leads to:
  - Keratinization of squamous epithelium of respiratory and GI tract mucous membranes
  - Metaplasia of salivary glands
  - Blunted choanal papillae
  - Widened choanal slit
  - Anorexia
  - Increased susceptibility to infection
  - Impaired ocular rod function/night blindness
  - Keratinization of conjunctiva/dry cornea/corneal abrasions
  - Possibly increased vocalizations (seen in cockatiels)
  - Possible change in feather pigmentation in some species

Vitamin D

- Cholecalciferol (vitamin D3) and ergosterol (vitamin D2)
- Birds synthesize cholecalciferol from cholesterol in skin exposed to ultraviolet (UV) light
- Most pet birds aren't exposed to enough UV light, so they a need dietary vitamin D source
- If birds are exposed to adequate UV light, they don't have a dietary requirement for vitamin D
- Without adequate UV light, requirement in chickens and turkeys is 200–1100 mg/kg diet
- As calcium (Ca) levels in diet decrease, phosphorus (P) levels increase, Ca:P ratio is affected and dietary requirement for vitamin D3 increases
- Birds cannot convert vitamin D2 into D3 as cats and dogs can
- Vitamin D3 acts as hormone to regulate Ca and P metabolism
- Essential in bone mineralization and eggshell formation
- Deficiency leads to:
  - Thin eggshells in hens
  - Osteomalacia
  - Pathologic bone fractures
- Excess can lead to toxicity with over-supplementation:
  - Ca is mobilized into blood and mineralizes soft tissues (joints, kidneys, heart, blood vessels, pancreas)
  - Documented with over-supplementation of hand-feeding formula and pelleted diets
  - Susceptibility varies by species, with macaws possibly more susceptible

Vitamin E

- D-tocopherol found in plant leaves
- Transported by plasma lipoproteins and incorporated into lipid of cell's plasma membranes
- Stabilizes membranes by reducing free radical peroxidation
- Works with superoxide dismutase, glutathione peroxidase, catalase to protect cell membranes
- Affects conversion of arachidonic acid into prostaglandins and thromboxanes, thereby altering immune response
- Dietary requirement varies depending on dietary levels of polyunsaturated fatty acids, vitamin A, betacarotene, and rancid fats, plus presence of other dietary antioxidants (i.e., selenium)
- Deficiency leads to:
  - Cell membrane dysfunction
  - Encephalomalacia ("crazy chick disease" = torticollis, vocalization, wing-flapping)
  - Muscular dystrophy
  - Ventricular myopathy
  - Exudative diathesis
  - Red blood cell fragility
- Excess can lead to deficiency of other fat-soluble vitamins

Vitamin K

- Has anti-hemorrhagic activity
- Ingested in plant leaves
- Cofactor of hepatic microsomal carboxylase that catalyzes carboxylation of glutamate in the function of osteocalcin, clotting factors VII/IX/X, and prothrombin
- Rapid turnover
- Not stored long in body
- Low requirement
  - Increases with infection, liver disease, vitamin K antagonists (antibiotics that deplete intestinal microflora producing vitamin K)

Omega Fatty Acids

Omega-6 and -3 polyunsaturated fatty acids (FA):

- Affect immune, cardiovascular, renal, and nervous system function
- Reduce inflammation and arthritis
- Omega-3 FAs produce fewer inflammatory cytokines than omega-6 FAs (arachidonic acid) that produce inflammatory prostaglandins, leukotrienes
- Most birds eating seeds and grains are not consuming a balance of polyunsaturated fatty acids but are getting more omega-6 FAs than omega-3 FAs
- Lack of omega FAs results in increased inflammation (itchy skin, unkempt feathers, poor feather quality)

Calcium

- Most important dietary mineral
- Needed for bone mineralization, eggshell calcification, nerve conduction, muscle contraction
- Requirement for psittacine bird species isn't known
  - Chickens require 0.1% of total diet for maintenance
  - Egg-laying chickens require 3.3% of total diet
  - Egg-laying budgerigars and cockatiels have been shown to require 0.85% and 0.35%, respectively, for normal egg calcification
  - African gray parrots may require more Ca than other psittacine bird species, as they more commonly develop hypocalcemic seizures
  - Requirement may be reduced when birds ingest larger amounts of pellets containing vitamin D3
  - Requirements may change with different life stages (i.e., increased with growth)
- Deficiency:
  - Occurs from:
    - Too little dietary Ca
    - Excess dietary P

        - Too little dietary vitamin D3
    - Is common in psittacine birds on all-seed/nut diets containing little Ca and vitamin D3
    - Leads to egg-binding in laying birds
- Excess:
    - One anecdotal report (D. Phalen 2005) of renal mineralization in budgerigars fed pellets exclusively
    - No evidence of renal mineralization noted in budgerigars fed all-pelleted diets in my 26-year career as an avian veterinarian recommending pellets to all psittacine bird species

Essential Amino Acids

- In chickens: arginine, leucine, isoleucine, lysine, methionine, phenylalanine, valine, tryptophan, threonine, and possibly glycine, histidine, proline because of insufficient synthesis relative to use
- Budgerigars also require glycine because they cannot synthesize enough
- Birds eating high-fat seed diets commonly cannot ingest enough seeds to meet their amino acid needs and eat until they are fat

Protein

- Feathers comprise the largest protein-containing portion of a bird's body
- Requirements increase with growth, reproduction, molting
- Required amino acids limit rate at which protein can be turned over and may require increased food ingestion to meet protein needs
- Theories about increased protein ingestion leading to gout in birds are not substantiated

Formulated Diets

- Nutritionally balanced products to be fed to companion birds include:
    - Pellets
    - Extruded diets
    - Pelleted material packed between intact (not ground up) grains and seed
- These diets are considered nutritionally balanced according to recommendations known for Galliformes and extrapolated according to AAV panel recommendations
- Few studies have been done determining nutritional needs of varied groups of psittacine bird species originating from diverse backgrounds
- Nutritional balance will change at different life stages (growth, reproduction, etc.)
- Pellets:
    - Mixture of ground grains with vitamins and minerals
    - Liquids are commonly added to mixture and then heated with dry or steam heat and forced through holes in a die with a roller
    - Pellets produced are of constant shape and size
- Extruded diets:
    - Commonly referred to as "pellets"
    - Mixture of ground grains with vitamins and minerals
    - Mixture is forced through an extruder with high heat and under pressure
    - Moisture is added as steam, which evaporates as mixture passes through extruder
    - Different-shaped holes in extruder plate determine the size and shape of food pieces created
    - High heat kills infectious organisms

FEEDING PRACTICES OF COMPANION PARROTS IN THE U.S.

General Practices

- Companion birds are not typically fed balanced diets
- Bird owners typically provide seed, produce, and a variety of other table foods, hoping birds will self-select nutritional balance from diverse options
- Typically, birds select out high-fat options and become obese
- To combat this trend, for the past 20+ years, avian veterinarians have recommended feeding formulated pelleted or extruded diets that are nutritionally balanced as the base diet

Problems with Feeding Formulated Diets

Bird owners often resist feeding these diets because:

- Birds don't readily transition to them
- Owners think of them as boring, lacking variety, and limiting opportunities for foraging
- Owners add produce and other table foods to formulated diets to increase variety and foraging

Pet Birds ≠ Wild Birds

Pet birds' nutritional requirements differ from those of wild birds:[2]

- Wild birds need calories to fly and to forage for food, mates, nest sites
- Wild birds graze all day without becoming obese, because they constantly burn calories
- Pet birds become obese because they sit in cages all day
- Pet birds are provided with food, so they don't have to hunt
- Pet birds bond to owners as "mates" and don't seek out other birds
- Food availability changes seasonally in the wild, so birds get more variety and balance
- Pet birds typically are fed the same foods every day, often in excess

Problems with Produce-Feeding

- Fruits and vegetables provide some micronutrients but are not the solution
- Produce is the basis of a healthy diet for relatively slow-growing, slowly reproducing humans
- Cannot provide adequate, balanced nutrition to meet the needs of rapidly growing, quickly reproducing pet birds
- Domestically grown produce has higher energy and water content and lower essential nutrient levels than wild counterparts
- Excessive produce-feeding may cause dietary nutrient dilution

Recommendations for Feeding Produce[3]

- Produce may be fed with a formulated diet to provide micronutrients, variety, and enrichment through foraging
- High water content of produce makes nutrient dilution minimal when fed as a complement to formulated diet
- Nutrient dilution is more common if formulated products make up <50% of total diet
- Table foods (including produce) should be no more than 25–30% of total diet so as not to upset nutritional balance

Seed/Nut Diets—A Persistent Problem

- Most common diet still fed to companion birds
- Are high in fat
- Lack fat-soluble vitamins (A, D, E, K) and critical minerals (calcium, phosphorus, sodium, zinc, iron, iodine, selenium, manganese)
- Are deficient in key amino acids (lysine, methionine, riboflavin, pantothenic acid, niacin, choline)
- Are overall nutritionally imbalanced
- Will be preferably consumed by pet birds if mixed with healthier food choices (selective eating)
- Veterinarians agree that pet birds' seed overconsumption causes many health problems:
    - Seed contains excess fat (cholesterol and triglycerides)
    - Seed lacks essential vitamins and minerals
    - Seed overconsumption leads to nutrient deficiencies
    - Lack of vitamin A → gout, poor skin/feather quality, respiratory problems
    - Lack of calcium, vitamin D → egg binding, soft-shelled/shell-less eggs
- Seed overconsumption leads to nutrient excesses
    - Excess fat → obesity, arthritis, high cholesterol, stroke, atherosclerosis, fatty liver/liver failure

The Bird Health and Nutrition Link

Diet Impacts Health

- Veterinarians, bird breeders, and bird-owning communities recognize the association between diet and health:[4,5]
    - Veterinarians treat many common bird health issues:
        - Reproductive disease/egg-laying problems
        - Liver disease
        - Kidney disease
        - Feather-picking/skin problems
    - Many health issues in birds may be due to poor nutrition:
        - Atherosclerosis/stroke
        - Obesity
        - Fatty liver disease
        - Kidney failure/gout
        - Egg-binding
        - Feather-picking/skin disease
        - Poor feather quality
- Veterinarians link many bird health issues to nutrient excesses and deficiencies:
    - Atherosclerosis/stroke—excess fat/cholesterol/triglycerides
    - Obesity—excess fat
    - Fatty liver disease—excess fat
    - Kidney failure/gout—lack of vitamin A
    - Egg-binding—lack of calcium, vitamin D
    - Feather-picking/skin problems—lack of vitamin A
    - Poor feather quality—lack of vitamin A, protein

Atherosclerosis[7-9]

- Accumulation of foam cells, lipids, calcium, and cellular debris leading to large fibrous-fatty plaques narrowing arterial lumen
- Extremely common in older, female Amazon, African gray, and male Quaker parrots
    - Amazon and African gray parrots may have increased incidence of atherosclerosis due to lack of exercise and consumption of high-fat diets
- Increases in total plasma cholesterol have been associated with increased incidence of atherosclerosis in psittacine species
    - Budgies fed a 2% cholesterol/15% lard diet developed hypercholesterolemia and significant atherosclerosis in 6 months
    - Quaker parrots fed a 1% cholesterol diet developed significant atherosclerosis in 2 months and advanced atherosclerosis in 4 months
        - Lesion severity and arterial cholesterol content were correlated with plasma cholesterol concentration

Supplementing Seed Isn't the Answer

- Many seed manufacturers mislead bird owners into thinking their products are nutritionally complete by coating seed hulls with supplements discarded by psittacine birds as they eat
- Well-intentioned owners offer a variety of healthy food choices with seeds and expect their pets to choose for proper balance
- When offered a varied diet, pet birds do not select for nutritional balance but select out what they like best (selective eating), which is typically the seeds
- Offering an abundance or a wide variety of food choices stimulates excessive reproductive activity, since food diversity signals breeding
- Homemade diets aren't the answer
    - Bird owners add foods to seed-based diets to try to provide variety and balance, but homemade combinations usually lack essential nutrients and contain excesses

Exotic Bird Ownership in the U.S.

Birds are Popular Pets

- American Pet Product Association 2019 survey of U.S. pets
    - Surveyed 25,109 households as a representation of total U.S. households
    - 5.7% of U.S. households owned pet birds = 6.9 million households
    - Average of 2.98 birds/household
    - 20.6 million pet birds in the U.S.
- ZuPreem 2014 survey of U.S. bird population
    - Surveyed 27,202 households as a representation of total U.S. households
    - 6.1% of U.S. households owned pet birds = 7.4 million households
    - Average of 2.50 birds/household
    - 18.6 million pet birds in the U.S.

ZuPreem 2014 Survey

- 1,680 bird caregivers with 4,200 birds of varying species
- Surveyed bird owners
- Divided birds into size categories
- Asked "What are bird owners looking for in a diet for their pets?"

- As expected, the response "provides complete and balanced nutrition" was the top food benefit sought by bird owners
- But the overwhelming response to "contains seed" was a surprise!
- Bird owners are not looking for sunflower seed-free foods
- Also not looking for "contains nuggets/pellets" at a high rate
- Owners who feed pellets/nuggets have concerns that birds won't eat it
- A significant number also don't think nuggets/pellets are nutritious enough!

- Conclusions:
  - More birds eat seeds and nuts than any other foods
  - Seeds and nuts are a greater percentage of U.S. birds' diets annually than any other foods
  - In terms of total food fed to birds in the U.S. in 2014, seed was, on average, 50% of their total consumption
  - 56% of birds in the U.S. eat no pellets
  - Of birds eating nuggets/pellets, this food made up only 31% of their total diets
  - Seeds and nuts are consumed by U.S. birds more than other foods, regardless of bird size
  - Seed makes up a larger part of small birds' diets and a smaller part of large birds' diets
  - If a bird eats pellets, it's only about 30% of any size bird's diet
  - Seeds make up a greater percentage of U.S. birds' annual diet, regardless of bird size

Pet Bird Feeding Facts

- Despite improvements in pet bird nutrition in recent years, birds in the U.S. still consume predominantly seed
- Most pet birds never see a veterinarian, so veterinarians never get the chance to improve most birds' diets
- Some nutritionally related conditions are less commonly seen in pet birds today because of better diets:
  - Sterile nasal granulomas in African gray parrots
  - Gout in all species
- Yet, many nutritionally related diseases persist in pet birds despite increased knowledge about pet birds' nutritional needs
- Statistics revealed in the ZuPreem survey explain why veterinarians continue to see these nutritionally based problems

How to Solve the Pet Bird Nutrition Problem

Formulated Products

- Are the solution as the base diet
- Provide all the "required" nutrients (including vitamins and minerals) without excessive fat, to promote birds' health
- Offer a manageable volume of food for a bird to consume daily that contains complete and balanced nutrition
- May be fed as base diet with smaller amounts of vegetables and fruit to provide micronutrients and enrichment through foraging
- May provide variety to birds' diets through different shapes, sizes, +/- colors without the risk of "selective eating"

Need to Create Awareness About Formulated Diets

- Veterinarians and technicians can continue to stress to bird owners the importance of these products and the detriment of seeds
- Formulated food manufacturers can continue to invest in educating bird owners about the health benefits of their products
- Pet stores can educate employees about the value of formulated products and the harm of seed-based diets
- Bird owners whose birds eat formulated diets can share their experiences

Difficulties with Seed to Pellet Conversion

- Veterinarians see a very small fraction of the total pet bird population
- Bird owners who never see veterinarians may not be aware of the importance of pelleted diets at all
- Many bird owners get frustrated or too stressed to continue conversion process if birds don't eat pellets immediately, and they just give up
- Veterinary staff can encourage bird owners to convert pets to pelleted diets, but owners need ongoing support during conversion process
- Bird owners who don't see veterinarians and try converting their birds to pellets need even more support and encouragement

ZuPreem Diet Conversion Project, 2018–19

Overview

- Goal is to provide support to veterinary staff and to bird owners as they convert pet birds off of seed-based diets and on to formulated diets
- Aim is to transition birds to healthier formulated products, regardless of brand or variety
- Veterinary staff examine pet birds to ensure they're healthy
- Veterinarians ask bird owners to follow one of three conversion methods or a variation of one
- ZuPreem aids veterinary staff in answering owners' questions during conversion process
- Owners are then sent follow-up surveys to see how conversion process went, what problems they encountered, what conversion methods worked best, etc.
- More than 175 veterinary hospitals from across U.S. and Canada enrolled
- Each hospital sees a minimum of 20 birds per week
- All birds must receive a veterinary checkup and be deemed healthy before conversion starts
- Project has been taking place for >1 year now
- All hospitals receive material to promote diet conversion:
  - Samples of pellets for large and small birds
  - Literature and links to online videos describing and showing conversion methods
- With 175 hospitals x minimum 20 birds/week x 50 weeks = >180,000 birds whose diets could be improved with pellets

ZuPreem Diet Conversion Methods

- ZuPreem website has detailed description of three conversion methods:
  - "Birdie's choice"
  - "Tough love"
  - "Slow and steady"
- Each method is broken down into a series of simple steps
- Each step has several tips to help owners implement
- Each step is described in words plus easy-to-follow illustrations
- Each method is also described in a real-life video
- Owner surveys:
  - Post-conversion surveys sent digitally or hard copy, depending on owners' preference, through veterinary hospitals to owners to gather info about conversion process
  - Owners fill out surveys and send back to hospitals
  - ZuPreem gathers survey results from hospitals
  - Statistics are run on conversion survey results

- Project results, presented at ExoticsCon 2019 by Dr. Alison Cummings,[10] will be published to help aid veterinary staff and bird owners in understanding how to make future diet conversion easier
- Survey questions:
  - What species is your bird?
  - How old is your bird?
  - Is your bird male or female, or is the gender unknown?
  - Were you able to convert your bird to eating pellets?
  - Which of the three methods did you use to convert your bird?
  - If another method, please describe
  - To what brand did your bird convert?
  - Which variety of that brand?
  - How long did it take for your bird to convert to eating pellets?
  - What problems did you encounter when you started the conversion process?

ZuPreem Diet Conversion Project Goals

- Identify which conversion methods work best for different species
- Determine average time for conversion of each species
- Determine which specific pellet brands and varieties worked best for particular species
- Identify most common problems owners encounter during conversion process and how to overcome them
- Set bird owners up for successful future pellet conversions

ZuPreem Diet Conversion Project Findings

- Conversion rate
  - Only 3/83 birds were reported to not convert (conversion rate of 96%)
  - Birds that did not convert followed the "slow and steady" method
    - Adult lovebird and juvenile green-cheeked conure
      - Both liked pellets, but owners were concerned they were not eating enough, so they continued to mix in seed
    - Juvenile budgerigar
      - Bird continued to eat around the pellets, so owner gave up
- Life stage vs. time to conversion
  - Regardless of life stage, 64% birds converted within 7 days and 90% within first month
  - Conversion within 3 days: adult 44%, geriatric 41%, juvenile 33%
  - Conversion within 7 days: juvenile 76%, geriatric and adult 59%
  - Geriatric group was the only group to have 100% conversion, with 94% converting in <1 month
- Time to conversion by method
  - Fastest method: Tough love with 90% of birds converting within 1–3 days
  - Most successful overall: Tough love with 100% of birds converting within the first month
  - Slowest method: Slow and steady, with 56% converting within 7 days, and 87% within the first month
  - Slow and steady was the only method that had birds not convert (5%)

ZuPreem Diet Conversion Project's Future Goals

- Follow birds before and after conversion to pelleted diet:
  - Body weight
  - Body condition score
  - Activity level
  - Blood lipid (cholesterol, triglyceride) concentrations
  - Overall health
- Hypothesize that birds on formulated diets will have more appropriate body weights/condition, improved activity level, more normal lipid concentrations, and overall better health than those on seed diets

Take-Home Messages

- Despite making great strides in pet bird nutrition over the past 20+ years, most bird owners are still ignorant about proper nutrition for their pets
- Pet birds are still predominantly eating high-fat, seed and nut diets and are not consuming balanced diets
- To get the proper balance of vitamins, minerals, amino acids, and other key nutrients, pet birds must consume a base diet of formulated pelleted diet
- The overwhelming majority of birds will convert to a pelleted diet from a seed-based diet
- For most birds, diet conversion takes less than 1 week regardless of life stage, conversion method, or bird family
- Conversion method was the only variable found to significantly affect conversion time when it comes to pet birds' conversion to pelleted diets
- The veterinary community—in association with pet stores, breeders, and bird food manufacturers—faces the ongoing task of better educating the bird-owning public about proper nutrition for companion birds

References

1. Orosz SE. Clinical avian nutrition. *Vet Clin Exot Anim.* 2014;17:397–413.

2. Perron F, Grosset C. The diet of adult psittacids: veterinarian and ethological approaches. *J Anim Physiol Anim Nutrition.* 2014;98:403–416.

3. Brightsmith DJ. Nutritional levels of diets fed to captive Amazon parrots: does mixing seed, produce, and pellets provide a healthy diet? *J Avian Med Surg.* 2012;26(3):149–160.

4. Shoemaker NJ, Lumeij JT, Beynen AC. Nutrition-related problems in pet birds. *Tijdschr Diergeneeskd.* 1999;124(2):39–43.

5. Koutsos E, Gelis S, Echols MS. Advancements in nutrition and nutritional therapy. In: Speer BL, ed. *Current Therapy in Avian Medicine and Surgery.* St. Louis, MO: Elsevier; 2016:142–176.

6. Beaufrere H. Atherosclerosis: comparative pathogenesis, lipoprotein metabolism, and avian and exotic companion mammal models. *J Exot Pet Med.* 2013;22(4):320–335.

7. Beaufrere H, Ammersbach M, Reavill D, *et al.* Prevalence of risk factors associated with atherosclerosis in psittacine birds. *J Am Vet Med Assoc.* 2013;242(12):1696–1704.

8. Beaufrere H, Cray C, Ammersbach M, *et al.* Association of plasma lipid levels with atherosclerosis prevalence in Psittaciformes. *J Avian Med Surg.* 2014;28(3):225–231.

9. Beaufrere H, Nevarez JG, Wakamatsu N, *et al.* Experimental diet-induced atherosclerosis in Quaker parrots (*Myiopsitta monachus*). *Vet Pathol.* 2013;50(6):1116–1126.

10. Cummings A. The pellet puzzle: an evaluation of three proposed pellet conversion methods. In: ExoticsCon Proc. 2019:148–149.

SPEAKER INFORMATION
(click the speaker's name to view other papers and abstracts submitted by this speaker)

Laurie Hess, DVM, DABVP (Avian Practice)
Veterinary Center for Birds & Exotics
Bedford Hills, NY, USA

URL: https://www.vin.com/doc/?id=9768894

Clinical Lipidology in Psittacine Birds

ExoticsCon Virtual 2020 Proceedings

Hugues Beaufrère, DVM, PhD, DACZM, DABVP (Avian), DECZM (Avian)
Department of Clinical Studies, Ontario Veterinary College, University of Guelph, Guelph, ON, Canada

Abstract

Lipids are a vast group of heterogenous molecular compounds that include fatty acids and their derivatives (such as triglycerides, sphingolipids, phospholipids, lipid mediators) or sterol compounds (such as cholesterol and esters, steroid hormones, fat-soluble vitamins, bile acids). They are also the main biomolecular constituents of plasma and occupy a central place in the pathophysiology of several common diseases of captive parrots. Two main groups of lipid disorders commonly seen in captive parrots are:

- Disorders characterized by excess lipid accumulation, such as obesity, hepatic lipidosis, atherosclerosis, lipomas, xanthomas, and hyperlipidemia
- Disorders related to reproductive diseases in female parrots, such as dysregulated vitellogenesis, egg yolk coelomitis, and emboli

While lipid disorders are common in captive parrots, especially pet parrots, diagnostic and therapeutic options are poorly validated and overall limited. Plasma cholesterol and triglyceride biochemical tests are the most commonly used clinically. Lipoprotein testing is limited because the analytical methods are either unpractical methods mainly used in research or unvalidated and inaccurate methods used in the clinical laboratory. As LDL is difficult to measure directly and the Friedewald formula is not accurate in parrots, it is recommended to only request HDL (for which analytical techniques are more robust across species) and non-HDL (which encompasses all atherogenic lipoproteins). Lipidomics is a new mass-spectrometry method that will likely revolutionize our understanding of lipid disorders in parrots, as it can measure hundreds to thousands of lipid species from minute plasma samples.

Therapy of lipid disorders mainly relies on correcting risk factors (diet, lifestyle) and using pharmacological interventions (GnRH agonists, statins, fibrates).

A Brief Overview of Lipid Diversity in Plasma

The diversity of biological lipid is staggering, and the reader is referred to the LIPID MAPS® website (https://lipidmaps.org) for the most up-to-date lipid classification and lipid species listing.

Lipids are predominant in psittacine plasma, as is the case in mammals. Lipids represent about 40% of plasma biomolecules in psittacine plasma (on a molar basis) with, for comparison, carbohydrates and proteins representing another 30% each. There are different types of lipids that have many functions.

Some lipids are in high concentration (measured in mmol/L or µmol/L) in the plasma and are part of what can be called the macrolipidome and typically include constitutive (such as cholesterol, cholesteryl esters, glycerophospholipids, and sphingolipids) or storage lipids (such as triacylglycerols). Their plasma concentrations get affected by a variety of metabolic disorders, and they also act as the precursors of many bioactive and signaling lipid molecules (most notable precursors are arachidonic acid and cholesterol, for instance).

On the other hand, numerous lipid species are bioactive molecules that act as mediators in various functions and are in very low concentration (measured in nmol/L or pmol/L, so a thousandth or millionth less concentrated than other lipids) and part of what can be called the microlipidome (or the mediator lipidome). The most important lipid mediators include inflammatory mediators (prostaglandins, leukotrienes, eicosanoids), prothrombotic molecules (thromboxanes), vitamins (vitamins A, D, E), pain mediators (endocannabinoids: anandamide and 2-AG), and steroid hormones (corticosterone, estradiol).

Most lipid species of the macrolipidome are transported within either lipoproteins (almost all lipids) or bound to albumin (free fatty acids), whereas lipid mediators are typically free in the plasma, bound to albumin, or transported by specific proteins.

Important plasma lipids of the lipidome are part of these five large groups of lipids:

Table 1. Important lipid groups found in parrot plasma

Lipid class	Subclass	Common lipid species in parrot plasma	Function
Sterol lipids	Cholesterol	Cholesterol	Cellular membranes
	Cholesteryl ester	CE (18:2)	Cellular membranes
	Bile acids	Taurochenodeoxycholic acid	Lipid digestion
	Steroid hormones	Corticosterone	Stress, sexual hormones, osmoregulation
Fatty acyls	Free fatty acids	Palmitic acid (16:0) Stearic acid (18:0) Oleic acid (18:1) Linoleic acid (18:2) Arachidonic acid (20:4)	Energy source Eicosanoid precursors
	Fatty acyl carnitine	Acetylcarnitine	Fatty acid mitochondrial transport
	Eicosanoids	Prostaglandin F2a Thromboxane B2	Inflammation, coagulation, egg laying

	Fatty amides	Anandamide	Endocannabinoid
Glycerolipids	Monoacylglycerols		Metabolite Endocannabinoid (2-AG)
	Diacylglycerols	DG (36:3)	Metabolite
	Triacylglycerols	TG (16:0, 18:1, 18:2) TG (16:0, 18:0, 18:3)	Energy storage
Glycerophospholipids	Glycerophosphates (PA)	PA (36:2)	Cellular membranes Signal transduction
	Phosphatidylcholines (PC)	PC (36:2), PC (34:2), PC (34:1)	
	Phosphatidylethanolamides (PE), serines (PS), inositol (PI), glycerol (PG)	PE (36:2), PG (36:2), PI (36:2), PS (21:0)	
	Lyso-PC, Lyso-PE, Lyso-PS, Lyso-PI	Lyso-PC (20:2)	Metabolites
Sphingolipids	Free sphingoid bases	Sphingosine (d18:0)	Metabolites, backbone of sphingolipids
	Ceramides	Cer (d18:1/22:0)	Metabolites, signaling molecules
	Sphingomyelins	SM (d18:1/16:0)	Cellular membranes, myelin
	Glycosphingolipids (cerebrosides, globosides, gangliosides)	Galactosyl(beta) ceramide (d18:1/24:0)	Cellular membranes, myelin

Sterol Lipids

These include free cholesterol, cholesteryl esters (CE), bile acids, and steroid hormones. Cholesterol and cholesteryl esters are the predominant lipid species in parrot plasma (about 50–60% on a molar basis). About one-third of total cholesterol is non-esterified (free) while two-thirds is esterified to various acyl chains, the most abundant by far being linoleic acid (18:2). CE (18:2) constitutes more than 70% of all cholesteryl esters. Cholesterol is synthetized in the liver (through a process involving the key enzyme HMG-CoA reductase, the target of statins), and exogenous sources of cholesterol are low in parrots as they are mainly frugivorous or granivorous animals (cholesterol is not present in plants). Cholesterol is mainly used for cellular membranes and as a precursor for other metabolites. Bile acids are cholesterol derivatives produced by the liver that assist in the emulsion and digestion of fat in the digestive system. The most common bile acid in parrot plasma is taurochenodeoxycholic acid and constitutes about 70% of all plasmatic bile acids. Bile acids are used as markers of liver health. Steroid hormones are synthesized from cholesterol and include hormones such as glucocorticoids (the main one being corticosterone in birds), mineralocorticoids (aldosterone), and sex hormones (produced by the gonads and adrenal glands) such as androgens (testosterone, androstenedione), estrogens (estradiol), and progestins (progesterone). They can affect the overall lipid metabolism and regulate vitellogenesis in hens.

Fatty Acyls

These include simple lipids such as free fatty acids and fatty acid derivatives. They typically have an even carbon chain in vertebrates. Fatty acids can be saturated or unsaturated (presence of double bonds in the hydrocarbon chain). The location of the double bond is important in terms of diet and health (omega-3 such as alpha-linolenic acid, DHA, EPA and omega-6 fatty acids such as linoleic acid). Linoleic (18:2) and alpha-linolenic acid (18:3) are essential fatty acids and must be supplied in the diet. Fatty acids are used as energy through mitochondrial beta-oxidation. Very long-chain PUFA (DHA, EPA, arachidonic acid) are the precursors of the eicosanoid lipid mediators such as prostaglandins, leukotrienes, thromboxanes and others. Non-esterified (free) fatty acids (NEFA) in the plasma come from either intestinal absorption or lipolysis (from triacylglycerols). To enter the beta-oxidation pathway occurring in the mitochondria, they are typically esterified to carnitine for transport through the mitochondrial membranes. Several acyl-carnitine species can be detected and quantified in parrot plasma. Beta-oxidation results in ketone bodies (with beta-hydroxybutyric acid being the most common in birds) that get elevated in the plasma when ketolysis processes are overwhelmed (such as with a large production of ketone bodies). The plasma concentration of NEFA can also be used clinically to assess the degree of lipolysis and negative energy balance. Eicosanoids are not routinely measured but can be useful to understand inflammatory pathways induced with various diseases. Fatty acyls represent only 5–10% of plasma lipids. The most preponderant fatty acid in the plasma is palmitic acid (16:0). In much lower concentrations, oleic acid (18:1) is the most common monounsaturated fatty acid, and linoleic acid (18:2) is the most common polyunsaturated fatty acid. Fatty acids produced by the preen gland are particular in that they are branched-chain fatty acids.

Glycerolipids

Glycerolipids include triacylglycerols (TAG, commonly called triglycerides in medicine even though it is an outdated term), diacylglycerols (DAG), and monoacylglycerols (MAG) and their alkylglycerol (glyceryl ethers) counterparts. They are composed of a molecule of glycerol esterified to one to three fatty acids in a specific order. The first fatty acid tends to be a saturated fatty acid (typically palmitic acid), the second a monounsaturated fatty acid (typically oleic acid), and the third is much more variable and is where polyunsaturated fatty acids are typically located. TAG composition typically reflects the diet as well as contribution of metabolism ("*de novo*" lipogenesis). "*De novo*" lipogenesis mainly occurs in the liver in birds with minimal contribution of the adipose tissues, unlike in mammals.[1] The main function of TAG is energy storage. Because of the large number of potential combinations of esterified fatty acids, glycerolipids species are very diverse in the plasma, and hundreds of distinct lipid species are present in the plasma. Glycerolipids constitute about 10% of plasma lipids on a molar basis, and TAG are the most abundant by far. TAG with 52 or 54 carbons—most being combinations of palmitic acid (16:0), stearic acid (18:0), oleic acid (18:1), linoleic acid (18:2), alpha-linolenic acid (18:3) and arachidonic acid (20:4)—are the predominant in parrot plasma. DAG and MAG are in much lower concentrations and are metabolic intermediates. A MAG of particular significance is 2-arachidonoyl-glycerol or 2-AG, which is the main endocannabinoid present in parrot plasma.

Glycerophospholipids

Glycerophospholipids (commonly abbreviated as phospholipids) are very diverse amphiphilic molecules that are made of a glycerol backbone with usually two esterified fatty acids (like glycerolipids) and an esterified phosphoryl part. The phosphoryl part can simply be phosphoric acid [glycerophosphates (PA)] or additional head groups to form more complex glycerophospholipid such as choline [glycerophosphocholine, also known as phosphatidylcholine (abbreviated PC) or lecithin], phosphatidylserines (PS), phosphatidylinositol (PI), phosphatidylglycerol (PG), phosphatidylethanolamines (PE) and others. The main function of phospholipids is structural as part of the cellular membranes. They can also function as signaling molecules (e.g., PIP2). The vast majority of phospholipids in plasma, both in number of lipid species and concentration, are the PC. Phospholipids are the second most abundant plasma lipid class in parrots after the sterols (around 14–15% of lipids). Lyso-PC (and other lysophospholipids) are metabolites generated by the loss of one of the two acyl chains. They are detected in the plasma too, but in much lower concentration. They are considered as metabolic intermediates.

Sphingolipids

Sphingolipids are incredibly diverse and complex lipids that are also structural lipids (membranes, myelin). They are in relatively low concentrations in the parrot plasma compared to other lipid classes (around 0.6%). While glycerolipids and phospholipids are based on glycerol, sphingolipids use long-chain sphingoid bases as backbones. Several sphingoid bases are possible, but sphingosine (abbreviated as d18:1) is the most common by far (as in mammals). When sphingosine is esterified to one fatty acyl chain, it is called a ceramide. Ceramides are intermediaries or byproducts of sphingolipid metabolism, but some may also serve as signaling molecules. On top of a fatty acyl chain (ceramides), other head groups may be found on the sphingosine backbone and determine the type of sphingolipid subclass such as the sphingomyelin (additional phosphocholine), sugars (cerebrosides, globosides, gangliosides). Sphingomyelins are the most abundant sphingolipids in parrot plasma.

A Brief Overview of Avian Lipoprotein Metabolism

Almost all lipids in the plasma are transported in the form of macromolecular aggregates of mixed lipid species and proteins known as lipoproteins. The exception is nonesterified fatty acids, which are transported bound to albumin. As avian lipoprotein metabolism has been comprehensively reviewed elsewhere,[1,2] only a brief recapitulative will be given here.

Lipoproteins are classified based on their density and size and are involved in different lipid transport pathways. They are of micellar lipid structure with a surface monolayer mainly composed of free cholesterol and phospholipids and a hydrophobic core composed of triacylglycerols and cholesteryl esters. Proteins called apolipoproteins are also present on the surface and have structural and signaling properties. Apolipoproteins are especially used for lipoprotein trafficking. Lipoproteins are classified based on their density into portomicrons, very low-density lipoproteins (VLDL), intermediate-density lipoproteins (IDL), low-density lipoproteins (LDL), and high-density lipoproteins (HDL).

Portomicrons are the avian equivalent to mammalian chylomicrons. The main differences are that portomicrons are not released into the lymphatic system (being poorly developed in birds) but in the portal vein. Portomicrons and VLDLs are very large lipoproteins that participate in the exogenous lipid pathway in birds, in which lipids absorbed through the intestinal walls are transported to the liver. They are therefore high in triglycerides. VLDLs and LDLs participate in the endogenous cholesterol/lipid pathway, in which cholesterol synthesized in the liver is transported into effector tissues. Two important enzymes that hydrolyze triacylglycerol within VLDL and IDL shrinking them into LDL at the effector sites are lipoprotein lipase and hepatic lipase. LDL is particularly high in cholesteryl esters. LDL particles are then picked up by the liver through LDL receptors for recycling. Apolipoprotein B100 is the main apolipoprotein of portomicrons, VLDL, and LDL. In humans, ApoB48 is present in chylomicrons but is not present in birds and is replaced by ApoB100.

HDLs participate in the reverse cholesterol transport pathway, in which cholesterol is brought back to the liver from effector tissues. Cholesterol is then recycled or excreted through the bile. HDLs are high in phospholipids and proteins and have apolipoprotein A as their main apolipoproteins. The plasma enzyme cholesteryl ester transfer protein (CETP) can move cholesterol from HDL to LDL in exchange for triglycerides. CETP has been measured in Quaker parrots and seems to increase with dyslipidemia.[3] In parrots, as in most birds and unlike humans, HDL is the main lipoprotein present in the plasma. Most of parrot HDL particles are small to medium in size, with a mean diameter of about 9.8 to 10.9 nm, much like in humans.[4] In female birds, another pathway (not present in mammals) is the egg lipid transport pathway, in which specialized, very low-density lipoproteins (the phospholipid-rich vitellogenin and triglyceride-rich VLDLy) transport lipid to the developing oocyte. It is under estrogenic influence. Vitellogenin and VLDLy have smaller diameters (around 30 nm) than regular VLDL (most being >31–36 nm) to be able to go through the granulosa basal lamina. These lipoproteins are not susceptible to lipoprotein lipase, unlike other lipoproteins.

Figure 1. Avian lipoprotein metabolism

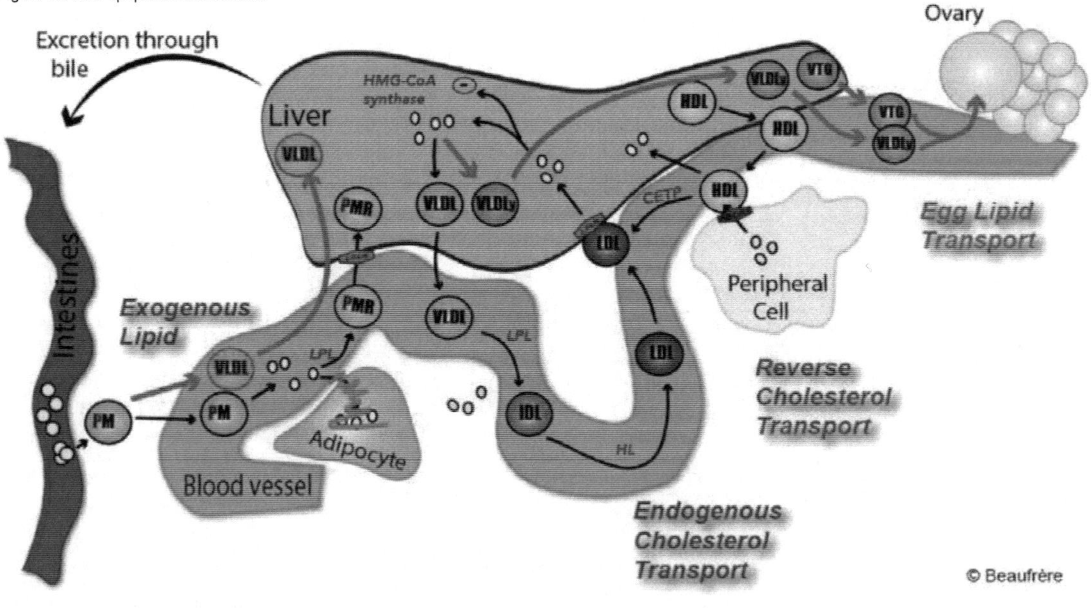

Lipidologic Diagnostic Methods

In terms of diagnostic tests and monitoring tools of fatty disorders, blood lipid analysis takes a center stage. Most if not all lipid-related disorders are either directly associated with dyslipidemic changes or secondarily through metabolic dysfunction and alterations of broad metabolic pathways. Most commonly used blood diagnostic tests in veterinary medicine and a fortiori in avian medicine tend to focus on plasma protein biomarkers, enzyme activities, electrolytes, or small metabolites instead of lipids. However, lipids are preponderant in the plasma, and lipidomic studies in parrots have shown that close to 1000–1500 distinct lipid species could be detected in Quaker parrot plasma. Consequently, lipids can serve as novel biomarkers for several diseases, mainly lipid-related and metabolic diseases. Unfortunately, there is a vast gap of knowledge for naturally occurring dyslipidemia and the association with common lipid-accumulation disorders in pet birds. Furthermore, most lipidologic diagnostic tests are not validated in Psittaciformes.

Lipidologic tests currently used and commercially available for birds include total cholesterol, total triglycerides, and lipoprotein wet-chemistry tests. Total cholesterol and triglycerides are the most useful tests to date in parrots and are expected to cross-react relatively well with tests developed for humans, as the cholesterol molecule is the same across species and the triglycerides test only targets the glycerol backbone of the molecule, which is also the same across species.

The cholesterol assay does not distinguish between free cholesterol and cholesteryl esters and in which lipoprotein the cholesterol is from, so it is a total cholesterol assay. Likewise, the triglycerides test does not distinguish between free glycerol, glyceryl ethers (DG-O), monoacylglycerols, diacylglycerols, and triacylglycerols. It is therefore more of a total glycerides test than a triacylglycerol test. Non-triacylglycerol glycerolipids account for only about 3–4% of plasma glycerolipids in parrots and could be considered negligible. However, in normal parrots, about 30% of what is measured as triglycerides is actually free glycerol (on a molar basis).[4] So it seems to be much more than in humans where free glycerol constitutes only 5–8% of total glycerides. This can interfere with clinical interpretation or the use of the triglycerides assay to estimate other values, such as VLDL or ratios. Pseudohypertriglyceridemia is caused by elevated blood free glycerol concentration in humans but has not yet been described in parrots. Point-of-care analyzers do not seem to perform well in parrots to measure total cholesterol or total triglycerides.[5,6]

As lipoproteins are the main lipid carriers and lipid distribution across lipoproteins varies, lipoprotein testing and profiling are routine in human medicine. The gold standard for lipoprotein analysis is separation by density gradient ultracentrifugation. This technique is extremely time consuming, requiring 6–30 hours of very high-speed centrifugation and specialized equipment.[7] It is not offered commercially for animals, but has been used for research in pigeons, chickens, quails, geese, Quaker parrots, and cockatiels.[1,8-14] While this method allows separation and isolation of lipoprotein classes and subclasses, it does not allow the determination of lipoprotein particle numbers and particle sizes. In order to analyze lipid contents of lipoprotein classes, a relatively large volume of blood is also needed, and each fraction must be carefully collected from the centrifugal tubes prior to analysis; therefore, ultracentrifugation of small volumes to obtain a lipoprotein panel, typically by analytical ultracentrifugation, can only allow the determination of relative distribution (% of total lipids). Absolute values are then determined from the total blood cholesterol levels and are not directly measured from each lipoprotein fraction.[9-11] Moreover, parrot-specific density gradient cut-offs have not been established.[9] Other techniques to measure lipoproteins include electrophoresis and nuclear magnetic resonance, but these techniques have not been described in parrots yet. Recently, a high-resolution technique using gel-permeation HPLC (LipoSEARCH®) has been reported to provide full lipoprotein profiling in parrots with a panel of 20 subclasses and particle sizes reported from just one drop of plasma.[4] Particle numbers can be obtained through NMR (not available because the algorithm to analyze the NMR spectrum is proprietary and based on humans) or the LipoSEARCH® panel. However, while these reference techniques are applicable to any animal species, they are not used or practical for clinical practice.

Avian lipoproteins present a large number of differences from mammals, and no lipoprotein diagnostic test has been validated in birds for routine use in practice. In addition, lipoproteins tend to degrade with any kind of storage, including in ultra-low freezers; therefore, it is best to perform lipoprotein analysis on fresh samples.[15] Direct LDL and HDL wet-chemistry analytes have been developed for human plasma and typically only measure their cholesterol content. These tests typically focus on HDL and use either selective precipitation or selective complexation methods. The Friedewald formula (LDL = cholesterol – HDL – VLDL/x with x = 5 in US units and x = 2.18 in SI units), an indirect method used most commonly for determination of LDL-C, is based on the assumption that the ratio between triglycerides and VLDL is the same across species. While being widely used, this formula is not accurate in Quaker parrots[4] and likely shows a similar lack of accuracy in other psittacine species. More studies are needed to determine a parrot-specific formula for indirect LDL estimation, but preliminary findings suggest using LDL = 0.75*non-HDL in healthy parrots.[4] The Friedewald formula is also known to be inaccurate with respect to hypertriglyceridemia in mammals, and avian vitellogenesis is expected to prevent its use in female birds.

Apolipoproteins (ApoB, ApoB/ApoA) are commonly measured, as they directly reflect lipoprotein particle numbers since there is only one apolipoprotein B molecule per non-HDL lipoprotein particle.[16] Birds lack Apo-E and ApoB-48. In addition, the parrot ApoB-100, the predominant apolipoprotein in portomicrons, VLDL, and LDL has only about 50% sequence homology with human Apo-B100. Therefore, apolipoprotein diagnostic tests based on human plasma are also unlikely to be useful in parrots.

Comprehensive lipid analysis by mass spectrometry, known as lipidomics, is rapidly revolutionizing the way metabolic disorders are investigated. Further, in the context of dyslipidemia, specific lipid species of complex lipids may be better targets or biomarkers than crude measurements of a single lipid molecule such as cholesterol and glycerol (for triglycerides). Thousands of species of triglycerides, phospholipids, and cholesteryl esters with a variety of saturated and unsaturated fatty acyl chains can be measured in plasma. Lipidomic analysis provides an important alternative approach that bypasses many limitations of conventional tests while providing considerably more information. The lipidome of male and female parrots also tends to be different. While it is not yet used on clinical patients, lipidomic research is underway in parrots (see presentations at this conference) and may lead to improved lipidologic diagnostic tests in the future. Specific biomarkers or lipidomic signature for specific diseases may be discovered with these techniques.

Until new tests are offered and validated, for now, total cholesterol and triglycerides remain the mainstay of reliable plasma lipidologic diagnostic tests in psittacine birds. Lipoprotein measurement using standard reference analyzers is likely not accurate, despite its use in multiple reference interval and experimental studies. If lipoproteins are still measured, they should be done on fresh plasma if possible and limiting the analysis to HDL-C without employing the Friedewald formula for LDL-C. Non-HDL-C can be used as a surrogate for VLDL/LDL cholesterol and encompasses all atherogenic lipoproteins. It has been shown to be an important risk factor for a number of diseases in humans and is now widely used in conjunction or replacement of LDL-C measurement. HDL measurement using standard reference analyzers uses a selective precipitation/complexation technique for non-HDL lipoproteins that may be more reliable across species than other lipoprotein analysis methods. However, this has not been studied in psittacine birds. Some other tests routinely available in the diagnostic laboratory, but seldom used, are beta-hydroxybutyrate (BHBA) and total non-esterified fatty acids (NEFA). BHBA is the main ketone in birds and is a marker of increased (and somewhat overwhelmed) fatty acid oxidation, whereas NEFA is a marker of lipolysis and negative energy balance. Both tend to get elevated in diabetes.

Another important point is that blood lipidologic tests are used to characterize dyslipidemia and risk factors for some diseases, but do not directly diagnose lipid-accumulation disorders such as atherosclerosis and hepatic lipidosis. While they may raise the suspicion for these disorders when biomarkers are elevated, other diagnostic tests should be used to diagnose these conditions (typically diagnostic imaging and tissue biopsies).

Lipid-Related Diseases

Dyslipidemia

Dyslipidemias (abnormal blood lipid concentrations) are extremely common in Psittaciformes and lead to a variety of primary and secondary disorders. The prevalence of hypercholesterolemia is estimated to be 20% in psittacine birds (n=5625, Beaufrère, unpublished, cut-off of 8 mmol/L). In addition, dyslipidemias are strongly associated with lipid-accumulation disorders in parrots, such as atherosclerosis, hepatic lipidosis, fatty tumors, obesity, as well as with chronic reproductive diseases in females.[9,17] Dyslipidemias have been poorly characterized other than for total hypercholesterolemia and hypertriglyceridemia. Likewise, spontaneous lipoprotein abnormalities are poorly characterized in pet birds.

Female birds undergoing vitellogenesis have an increased cholesterol and triglyceride level in the plasma. Their lipids are within vitellogenin and VLDLy, which are resistant to the action of lipoprotein lipase. This means that in dysregulated vitellogenesis, the dyslipidemia may be more long lasting and harder to control.

Diet undoubtedly plays an important role in the emergence of dyslipidemia either through increased calorie or fat intake and through multiple deficiencies. A moderate increase in fructose intake has not been associated with dyslipidemia in zoo macaws (Desmarchelier, unpublished). A seed-based diet may lead to chronic dyslipidemia and nutritional deficiencies. For instance, sunflower seeds are composed of more than 50% fat even if its fatty acid profile is good.

Results of studies on cholesterol in parrots have been inconsistent in terms of sex difference, with females being found to have higher HDL than males in some studies, higher LDL, or no difference at all (most commonly). Other factors may also affect blood lipid interpretation, such as postprandial hyperlipidemia, aging, and hepatobiliary diseases that may cause an elevation in cholesterol subsequent to cholestasis. Hypothyroidism is also a well-known cause of dyslipidemia but is rare in pet birds. In psittacine birds reported to have spontaneous or experimental hypothyroidism or goiter, hypercholesterolemia was frequently encountered.

Clinical experience shows that both non-HDL and HDL increase in dyslipidemia, but the ratio HDL/non-HDL tends to shift. Some species tend to have higher blood lipid values than others, in particular Quaker parrots and Amazon parrots. Other risk factors include the diet and lack of activity. The normal diet of most parrots should not include animal products, which are rich in saturated fat and cholesterol (the latter not being present in plants). Strong dyslipidemia has also been experimentally induced in Quaker parrots fed a 1% cholesterol pelletized diet.[3] In the same species, a seed diet led to elevation in triglycerides and lipemia, and overweight birds had higher cholesterol, triglycerides, lipemia, and non-HDL cholesterol.[9]

Hepatic Lipidosis

Hepatic lipidosis is characterized by the accumulation of triglycerides in hepatocytes, which may progress to liver dysfunction when severe. The prevalence seems higher in some species such as Amazon parrots and particularly in Quaker parrots with a reported prevalence of 21%.[18] The background prevalence in psittacine species seems to be around 6%.[18] The pathophysiology is poorly understood in pet birds and seems to be associated with high-fat diet. Possible etiology includes increased fatty acid uptake (from food or adipose tissue lipolysis), increased *de novo* synthesis (from carbohydrates or vitellogenesis), decreased fatty acid degradation (from impaired or saturated mitochondrial beta-oxidation), or decreased hepatic excretion (impaired VLDL excretion). In laying chicken, the disease is strongly associated with estrogen-induced lipogenesis in the liver, so it could be a risk factor in some female parrots with reproductive disorders as well. However, in Quaker parrots, the disorder is more common in males, and no sex predisposition was found in a group including Amazon parrots, cockatoos, cockatiels, macaws, and African grey parrots.[18] Hepatic lipidosis is associated with dyslipidemic changes, mainly hypertriglyceridemia, in most species. However, this has not been well demonstrated in parrots. Standard hepatic biomarkers may also be useful, in particular bile acids, but they tend to have low sensitivity except in severe disease. Radiographs and ultrasound may show hepatomegaly and a hyperechoic liver, respectively, but cannot confirm the diagnosis. On the other side, CT scan will confirm the increased fat content of the liver (with low HU). While a diagnosis of fatty liver can be made on CT alone, the histologic characterization will not be known, such as the degree of fibrosis and inflammation. Hepatic biopsies are relatively easy to take in birds via coelioscopy, either through a midline approach into the ventral hepato-peritoneal cavities or through a lateral approach through the air sacs.

Atherosclerosis

Atherosclerosis is an inflammatory and degenerative disease of the arterial wall characterized by disorganization of the arterial intima from the accumulation of inflammatory cells, fat, cholesterol, calcium, cellular debris, and inflammatory cells and potentially leading to complications such as stenosis, ischemia, thrombosis, hemorrhage, and aneurysm. Atherosclerosis is probably an underlying lesion in the majority of noninfectious cardiovascular diseases diagnosed in pet birds and is undoubtedly the most common lesion of the cardiovascular system. The centrally located arteries (brachiocephalic trunk, ascending aorta, pulmonary arteries) are most often affected, but lesions can be found in any artery. Known and suspected risk factors include age, female sex, species (Amazon parrots, African grey parrots, cockatiels), increased plasma total cholesterol and triglycerides, high-energy and high-fat diet, and physical inactivity.[17] However, in Quaker parrots, males are predisposed.[18] Clinical signs are uncommon in psittacine atherosclerosis but, when present, consist of sudden death, congestive heart failure, dyspnea, neurologic signs, respiratory signs, exercise intolerance, and ataxia. Atherosclerotic disease in parrots is caused by arterial luminal stenosis and rarely by atherothrombosis (clots). While atherosclerosis and hepatic lipidosis share some risk factors, no statistical associations have been found between the two disorders in epidemiological studies (contrary to popular belief).[18,19]

The diagnosis of atherosclerosis mainly relies on imaging to detect arterial calcification, strongly associated with the lesions. A subjective increased arterial size on x-rays or CT scan is not reliable to diagnose atherosclerotic lesions. Arteries are not accessible to ultrasound evaluation, as they are suspended between air sacs. Blood lipid diagnostics only assess risk factors.

Fatty Tumors

Fatty tumors are frequently diagnosed in captive psittacine birds and include lipoma, liposarcoma, myelolipoma, and xanthoma (a pseudotumor). The prevalence of lipomas is fairly high, and they are some of the most common neoplasia of parrots. In Quaker parrots, the prevalence for lipoma/liposarcoma was 1.9% on necropsy and 11.5% on biopsy samples.[18] The prevalence of xanthomas was 0.7 and 4.9%, respectively. Other epidemiological studies in other parrot species confirmed that these tumors were the most commonly encountered neoplasia.

Reproductive-Associated Lipid Disorders

Psittacine birds have a high prevalence of female reproductive-associated disorders, especially in smaller species such as cockatiels, lovebirds, and budgerigars. Most of these conditions are associated with unregulated vitellogenesis or lesions resulting from egg yolk deposition or embolization. Addressing the underlying issue will most often correct the associated dyslipidemia.

Other Less Common Disorders

Lipids are also involved in numerous metabolic pathways, endocrine diseases and nutritional diseases, which will consequently lead to secondary lipid disorders and dyslipidemic changes. In addition, a constellation of other less prevalent lipid disorders has been reported in various species of birds including parrots, such as corneal lipid deposition, endogenous lipid pneumonia, lipid storage disorders (lysosomal storage disease), and lipodystrophies. Lysosomal storage disease is uncommonly reported in parrots, with a single case in an African grey parrot with a massive spleen, suspected of having Niemann-Pick disease (accumulation of sphingomyelins).

Therapy

For most lipid-accumulation disorders, dyslipidemia is a strong contributing factor, thus treating dyslipidemia is strongly recommended. Likewise, blood lipid diagnostics can be used to monitor therapeutic progress. Dietary improvement and lifestyle changes are the easiest ways to manage most of these disorders. Having primarily a pelletized diet supplemented with fresh fruits and vegetables as well as discontinuing animal products will help. Low-fat pellets are also commercially available. Omega-3 fatty acids as found in fish and flaxseed oil have shown some positive effects on blood lipid in studies in psittacine birds.[11] Increased exercise was associated with weight loss and increased HDL cholesterol in laboratory Amazon parrots.[20]

In female birds, vitellogenesis and reproductive disorders should first be addressed by decreasing reproductive stimuli and by using GnRH-agonist therapy (deslorelin implants).

The main family of hypolipidemic drug that has been studied and used in Psittaciformes is the statins. They are cholesterol inhibitors that target the rate-limiting enzyme HMG-Co reductase in the liver. Rosuvastatin at 10 mg/kg PO q 12 h did not reach therapeutic plasma concentration in Hispaniolan Amazon parrots. Atorvastatin at 10 mg/kg PO q 24 h in hypercholesterolemic Hispaniolan Amazon parrots led to decreased plasma total cholesterol, but this change was not statistically significant.

Atorvastatin at 10 mg/kg PO q 12 h is preferred by the author and seems safe. Scientific information is still limited, but preliminary findings suggest that some hypolipidemic effect can be seen in parrots. Statins also have anti-atherosclerotic effects and beneficial effects in hepatic lipidosis that go beyond their primary hypocholesterolemic effect. Fibrates (e.g., gemfibrozil) are hypotriglyceridemic agents that can also be used, but there is no data on their use in Psittaciformes. Statins and fibrates should not be combined, and both may lead to muscular adverse effects. Grapefruit should not be given to birds on statins, as it may drastically increase plasma concentrations and side effects. It is linked to the presence of furanocoumarin, a cytochrome P450 inhibitor, which is particularly high in grapefruit.

Acknowledgement

We acknowledge the support of Rolf C. Hagen Inc. to the University of Guelph research parrot colony.

Disclosure Statement

Some of the research results summarized in this handout were obtained from funding from the Morris Animal Foundation (grant ID: D19ZO-301) and the OVC Pet Trust.

References

1. Alvarenga RR, Zangeronimo MG, Pereira LJ, Rodrigues PB, Gomide EM. Lipoprotein metabolism in poultry. *World Poult Sci J.* 2011;67:431–440.

2. Beaufrere H. Atherosclerosis: comparative pathogenesis, lipoprotein metabolism and avian and exotic companion mammal models. *J Exot Pet Med.* 2013;22(4):320–335.

3. Beaufrère H, Nevarez JGG, Wakamatsu N, Clubb S, Cray C, Tully TNN. Experimental diet-induced atherosclerosis in Quaker parrots (*Myiopsitta monachus*). *Vet Pathol.*

2013;50(6):1116–1126.

4. Beaufrere H, Gardhouse S, Ammersbach M. Lipoprotein characterization in Quaker parrots (*Myiopsitta monachus*) using gel-permeation high-performance liquid chromatography. *Vet Clin Path*. In press.

5. Barboza T, Beaufrère H. Comparison of a point-of-care cholesterol meter with a reference laboratory analyzer in companion Psittaciformes. *J Avian Med Surg*. 2019;33(1):7.

6. Irvine KL, Mans C, Friedrichs KR. Method comparison using 2 point-of-care meters and a reference analyzer for measuring blood triglycerides in psittacine birds. *J Avian Med Surg*. 2019;33(3):229.

7. Caslake M, Packard C. The use of ultracentrifugation for the separation of lipoproteins. In: Rifal N, Warnick G, Dominiczak M, eds. *Handbook of Lipoprotein Testing*. 2nd ed. Washington DC: AACC Press; 2000:625–646.

8. Langelier M, Connelly P, Subbiah MT. Plasma lipoprotein profile and composition in White Carneau and Show Racer breeds of pigeons. *Can J Biochem*. 1976;54(1):27–31. www.ncbi.nlm.nih.gov/pubmed/175901. Accessed August 28, 2012.

9. Belcher C, Heatley J Jill, Petzinger C, *et al*. Evaluation of plasma cholesterol, triglyceride, and lipid density profiles in captive monk parakeets (*Myiopsitta monachus*). *J Exot Pet Med*. 2014;23(1):71–78.

10. Petzinger C, Larner C, Heatley JJ, Bailey CA, MacFarlane RD, Bauer JE. Conversion of α-linolenic acid to long-chain omega-3 fatty acid derivatives and alterations of HDL density subfractions and plasma lipids with dietary polyunsaturated fatty acids in Monk parrots (*Myiopsitta monachus*). *J Anim Physiol Anim Nutr (Berl)*. 2014;98(2):262–270.

11. Heinze CR, Hawkins MG, Gillies LA, *et al*. Effect of dietary omega-3 fatty acids on red blood cell lipid composition and plasma metabolites in the cockatiel, *Nymphicus hollandicus*. *J Anim Sci*. 2012;90(9):3068–3079.

12. Oku H, Ishikawa M, Nagata J, Toda T, Chinen I. Lipoprotein and apoprotein profile of Japanese quail. *Biochem Biophys Acta*. 1993;1167:22–28.

13. Barakat HA, St Clair RW. Characterization of plasma lipoproteins of grain- and cholesterol-fed White Carneau and Show Racer pigeons. *J Lipid Res*. 1985;26(10):1252–1268.

14. Hermier D, Saadoun A, Salichon MR, Sellier N, Rousselot-Paillet D, Chapman MJ. Plasma lipoproteins and liver lipids in two breeds of geese with different susceptibility to hepatic steatosis: changes induced by development and force-feeding. *Lipids*. 1991;26(5):331–339.

15. Zivkovic AM, Wiest MM, Nguyen UT, Davis R, Watkins SM, German JB. Effects of sample handling and storage on quantitative lipid analysis in human serum. *Metabolomics*. 2009;5(4):507–516.

16. Okazaki M, Yamashita S. Recent advances in analytical methods on lipoprotein subclasses: calculation of particle numbers from lipid levels by gel permeation HPLC using "spherical particle model." *J Oleo Sci*. 2016;65(4):265–282.

17. Beaufrere H. Avian atherosclerosis: parrots and beyond. *J Exot Pet Med*. 2013;22(4):336–347.

18. Beaufrère H, Reavill D, Heatley J, Susta L. Lipid-related lesions in Quaker Parrots (*Myiopsitta monachus*). *Vet Pathol*. 2019;56(2):282–288.

19. Beaufrère H, Ammersbach M, Reavill DR, *et al*. Prevalence of and risk factors associated with atherosclerosis in psittacine birds. *J Am Vet Med Assoc*. 2013;242(12):1696–1704.

20. Gustavsen KA, Stanhope KL, Lin AS, Graham JL, Havel PJ, Paul-Murphy JR. Effects of exercise on the plasma lipid profile in Hispaniolan Amazon parrots (*Amazona ventralis*) with naturally occurring hypercholesterolemia. *J Zoo Wildl Med*. 2016;47(3):760–769.

Author's note: References were limited to 20. Additional references are available upon request.

SPEAKER INFORMATION
(click the speaker's name to view other papers and abstracts submitted by this speaker)

Hugues Beaufrère, DVM, PhD, DACZM, DABVP (Avian), DECZM (Avian)
Department of Clinical Studies
Ontario Veterinary College
University of Guelph
Guelph, ON, Canada

URL: https://www.vin.com/doc/?id=9768893

Introduction to Radiography of Raptors and Companion Birds

ExoticsCon Virtual 2020 Proceedings

Seth C. Oster, DVM

Department of Clinical Sciences and the Southeastern Raptor Center, College of Veterinary Medicine, Auburn University, Auburn, AL, USA

Abstract

Raptor and companion bird radiology presents several unique challenges when compared to mammalian counterparts. This masterclass is designed to give the practitioner the basic tools needed to properly acquire diagnostic radiographs and to interpret these images. We will discuss proper restraint options, positioning, and image acquisition techniques. Following this, we will review several examples of normal and abnormal radiographic anatomy.

Restraint

Restraint of birds for radiographic examination is commonly achieved through sedation or anesthesia.[1] Conscious restraint is not ideal in most cases and increases the risk of injury for the bird and decreases the ability to position most birds correctly for radiographs. This author prefers gas anesthesia with isoflurane due to its rapid onset and recovery. Sedation, with compounds such as midazolam, has been well tolerated by most birds. If the condition is painful, often sedation alone is not enough to allow for adequate relaxation for positioning.

Positioning

Positioning is key to diagnostic radiographs. There are a few items that can make positioning easier. Foam wedges or pads can be very useful to elevate heads or hips to get proper alignment for lateral views. Foam can also be useful in separating legs in a lateral view. White tape or masking tape can be used to secure limbs in position with minimal damage to feathers.[2] Bird Boards have also been used to aid in positioning.[3] These devices are made of acrylic and are more appropriate for larger birds. Most of these devices have straps or ties to secure wings and legs and an acrylic plate to secure the head in place. This author prefers the use of foam and tape.

A properly positioned ventrodorsal view of the coelom should, if possible, include the thoracic inlet cranially and legs caudally. The patient's dorsum should be laid flat on the table with the keel pointed straight toward the ceiling. This should result in the keel being superimposed directly over the thoracic vertebrae in the resulting imaging.[1] Both legs should be extended equally to prevent rotation of the body. Wings should also be extended equally to prevent rotation of the body. When using tape to restrain, wing tape should touch the wing over the distal radius/ulna. Leg tape should contact the leg over the distal tarsometatarsus. The keel should be aligned with the "+" created with the light of the beam, and the beam should be centered over the caudal portion of the keel. The entire coelomic cavity should fit within a single image in all but the largest bird. Labeling with a metal R or L is, of course, part of standard procedure.

When positioning for a lateral view of the coelomic cavity, both wings should be fully extended dorsally. It is important to not superimpose the wings directly on top of each other. The down wing is commonly placed more cranial to allow easy identification on the resulting image. Tape can again be placed over the distal radius/ulna. In large birds, a foam pad or wedge between the wings can prevent over-extending wings. Legs should be extended ventrocaudally. Again, avoid superimposing legs by moving the down leg more cranially. Tape can again be placed around the distal tarsometatarsus. In most birds, the hips are narrower than the shoulders. This will cause most birds to rotate their hips toward the down side, twisting the coelom. Placing a small foam wedge under the hips will help prevent this rotation and make a well-aligned image. To confirm correct positioning, the heads of the humerus should be superimposed as well as the head of the femur on the final image.

Lateral images of the wings are designed to have the beam base from the ventral aspect of the wing through the dorsal aspect. To position this view, place the patient on the table in dorsal recumbency. Extend the wing that you wish to view until it reaches the natural end of its range of motion. In this position, it is best to use tape passing over the metacarpus to make sure both the elbow and carpus are fully extended. In larger birds, tape can be added over the mid radius and ulna to spread the pressure of the tape out. Center the x-ray tube over the mid radius and ulna if you can fit the entire wing in the field. In larger birds, it may take two or three images to get the entire wing radiographed.

The caudal-cranial view of the wing is another often-used position. This positioning starts with the patient in dorsal recumbency on the table. Feet should be toward the restrainer, and the head should point away from the restrainer. While keeping the patient's head flat on the table, the restrainer grasps the legs and tail in a glove hand and raises them toward the ceiling. Be aware this technique has an increased risk of regurgitation. The other hand holds the primary feathers of the wing and extends the wing out from the body. The beam should pass from the caudal aspect of the wing through the cranial aspect. Be sure to avoid getting fingers in the view. The beam can be centered over the proximal radius and ulna. In large birds, multiple views may be required to get the entire wing imaged. If the image is properly positioned, the radius and ulna will be superimposed in the resulting image.

Lateral images of the legs can be accomplished by placing the patient in the same position as the lateral whole-body images. In the case of the legs, the down leg should be moved cranially (example, the right lateral view of the legs should have the right leg moved cranially). The toes should be spread out and taped in position to prevent superimposition. The beam should be centered over the proximal tibiotarsus, and the entire leg from the femur to the toes should be included in the image if possible. Large birds or long-legged birds may require multiple images.

A ventrodorsal view of the pelvis is achieved by placing the patient in dorsal recumbency. The legs are fully extended and tapped in place, similar to the positioning for the VD images of the whole body. Toes should be spread out and tapped in place to prevent superimposition. The beam should be centered over the proximal tibiotarsus, and the entire leg from the femur to the toes should be included. In larger birds or long-legged birds, multiple images may be required.

The H view is a recent addition to the toolkit of avian practitioners. It is used to examine the coracoid, clavicle, and scapula without these bones being superimposed upon one another. This image is acquired by placing the patient in the same position that would be used for a standard whole-body ventrodorsal view. It is important that the patient's head and feet are aligned with the length of the table. This technique requires a machine that will allow for the head of the x-ray machine to rotate to 30–45°. The exact angle depends on the species of bird being examined. This practitioner routinely starts at 30° for most patients and increases the angle if there is still superimposition of the bones of interest. Rotate the head of the machine to the desired angle and then move the machine caudally in relation to the patient. The beam should be aimed at the shoulders of the patient passing through it in a caudal-ventral/cranial-dorsal angle. As this beam is at an angle to the patient, traditional measurement for the determination of technique can be difficult. Technique set based on VD measurements of the patient tend to result in overexposed images. This practitioner uses a VD measurement and decreases the technique by 1 cm in most cases.

Technique

Radiographic technique can be highly individual for each machine. It is recommended to develop your own technique chart for avian radiographs. In most avian[4] patients, a low milliamperage-sec (mAs) and kilovoltage peak (kVp) are used. A suggested technique chart is provided below.

Suggested technique chart

CM	kVp	mAs
1	50	5

2	52	5
3	54	5
4	56	5
5	58	5
6	60	5
7	62	5
8	64	5
9	66	5
10	68	5
11	70	5
12	72	5
13	74	5
14	76	5
15	78	5

It is important to consider conventional film vs. digital systems when determining technique. Generally, digital systems require slightly higher mAs and kVp than conventional film systems (around 10 to 15%). All digital systems are not created equal. Digital systems use an algorithm to create the image after it is acquired on the cassette or plate. The algorithms vary based on what you set the computer to (for example, a thorax algorithm reads the plate differently than an abdominal algorithm). To be clear, the algorithm is a processing function of the computer; it is not the mAs or kVp that you set your radiograph machine to. These algorithms can vary between manufacturers. Some systems have exotic/avian algorithms. For those without, use a setting for a small mammal (such as a cat) extremity. Consult your manufacturer if you are having trouble with your image quality.

Image Interpretation

Image interpretation should be done in a systematic manner. This author prefers to first evaluate position and technique to confirm they are adequate. Next, examination of the skeletal structures, noting any abnormalities. Following that, this author then examines soft tissue structures. There can be a significant amount of variation in size and position of organs between different species of birds. A reference text or reference films are important tools to have on hand. Several examples of raptor and psittacine anatomy will be covered in this masterclass.

References

1. Naldo J, Saggese M. *Avian Medicine.* St. Louis, MO: Elsevier; 2016.
2. Krautwald-Junghanns M, Pees M. *Handbook of Avian Medicine.* St. Louis, MO: Elsevier; 2009.
3. Krautwald-Junghanns M. *Essentials of Avian Medicine and Surgery.* Ames, IA: Blackwell; 1997.
4. Silverman S, Tell L. *Radiology of Birds.* St. Louis, MO: Elsevier; 2010.

Speaker Information

(click the speaker's name to view other papers and abstracts submitted by this speaker)

Seth C. Oster, DVM
Department of Clinical Sciences & the Southeastern Raptor Center
College of Veterinary Medicine
Auburn University
Auburn, AL, USA

URL: https://www.vin.com/doc/?id=9768896

Sedation in Birds

ExoticsCon Virtual 2020 Proceedings

Christoph Mans, DrMedVet, DACZM, DECZM (ZHM); Grayson A. Doss, DVM, DACZM
School of Veterinary Medicine, University of Wisconsin-Madison, Madison, WI, USA

Introduction

Sedation of dogs and cats in veterinary practice is daily routine for a variety of procedures, such as radiographs and ultrasonography or other nonpainful, but potentially stressful, procedures. However, historically for avian patients, either manual restraint of conscious birds or general anesthesia is typically performed to complete most clinical procedures. General anesthesia predisposes birds to cardiovascular and respiratory depression and may cause aspiration of gastric or crop contents and hypothermia. In contrast, manual restraint in conscious birds is simple to perform but can have negative consequences, including stress to the bird and/or handler, negative conditioning to the clinic environment (e.g., the person restraining or the towel used for restraint), hyperthermia, and the predisposition of trauma to the handler and/or bird. Several recent studies demonstrated that manual restraint of birds causes increased body temperature and respiratory rate.[1,2] In sick, old, or very stressed birds, acute collapse and death secondary to manual restraint have been reported. Therefore, sedation techniques provide a useful alternative for reducing physiologic stress in birds undergoing nonpainful clinical procedures. Further, sedation in birds provides easier restraint and increases the safety of many clinical procedures (e.g., blood collection, radiography, ultrasonography) and allows for a more complete examination, which would otherwise only be achieved under general anesthesia. Using safe and effective sedative protocols in pet birds provides substantial benefits to the patients as well as the veterinarian and staff, and should be considered for a variety of clinical procedures.

Route of Drug Administration

Historically, sedative drugs have been most commonly administered by intramuscular injection in birds. Intranasal administration of sedative drugs in birds has gained increased attention, and most sedation studies performed in psittacines have used the intranasal route for drug administration. Intranasal administration of midazolam in Hispaniolan Amazon parrots, ring-necked parakeets, budgerigars, and canary finches resulted in rapid onset of dose-dependent degree sedation, which is completely reversible with intranasal flumazenil.[2-5] Intranasal drug administration offers an alternative, noninvasive technique for drug administration in birds (Table 1). It is characterized by its ease of administration, high bioavailability, rapid onset of action, and reduced pain compared to intramuscular administration. Elevation of muscle enzymes in biochemistry panels secondary to intramuscular drug administration is avoided if intranasal administration is used instead. In addition, clients perceive the intranasal route as noninvasive compared to intramuscular injection, which leads to better client compliance in cases in which sedation is recommended. The time to onset of sedation is rapid, typically within 3–5 minutes.[2-5] However, limitations of intranasal administration include incomplete drug delivery due to sneezing during administration, physiologically narrowed nostrils (e.g., cockatoos), or upper respiratory disease (e.g., blocked or stenotic nostrils). In some cases in larger birds (e.g., macaws), the drug volume can also limit the effectiveness and produce excessive sneezing, therefore leading to incomplete drug delivery. Higher concentrated drugs (e.g., midazolam 50 mg/ml, ZooPharm, Windsor, CO) are available, but intramuscular administration might be more feasible in these cases.

Table 1. Comparison of intramuscular and intranasal drug administration in pet birds

	Intramuscular administration	Intranasal administration
Pro	Reliable drug delivery Faster administration	Noninvasive and not painful Higher client acceptance Reduced risk of needle-stick injury No changes in the biochemical panel
Con	Invasive Potentially painful Possible post-injection hemorrhage Changes in the biochemical panel	Risk of incomplete drug delivery Difficult to administer larger volumes May cause sneezing More difficult in birds with feathered nares

Drugs Used for Sedation

Midazolam

Midazolam is currently the most commonly used drug for sedation of pet birds and has a wide margin of safety. Midazolam has sedative, muscle-relaxing, anxiolytic, amnestic, and appetite-stimulating properties in birds.[2,6,7] The injectable form of midazolam (midazolam hydrochloride, 5 mg/ml; Hospira Inc., Lake Forest, IL) or a more concentrated form (50 mg/ml; ZooPharm, Windsor, CO) can be administered intranasally and intramuscularly without side effects.[2-4] Dosages commonly used in pet birds range from 0.5–3 mg/kg.[8] At the University of Wisconsin, we routinely use 2 mg/kg of midazolam in pet birds, if administered intranasally and as the sole sedative agent. In smaller birds such as finches or budgerigars, we routinely use 4–6 mg/kg of midazolam if administered alone. Doses should be adjusted accordingly in debilitated patients to avoid adverse effects.

Diazepam

Diazepam is of similar efficacy as midazolam in birds following intranasal administration, but it has a longer onset time and duration of action.[4,5,9] While less commonly used for intranasal administration in birds, diazepam represents a suitable alternative in cases in which midazolam might not be available. The intramuscular administration of diazepam should be avoided due to delayed absorption and muscle irritation.[5] Dosages commonly used in pet birds range from 0.2–2 mg/kg if used as a sole sedative agent. Dosages as high as 10–15 mg/kg have been administered to finches and budgerigars without significant side effects.[4,5,9]

Butorphanol

Butorphanol is currently the most commonly used opioid analgesic in birds. Besides its analgesic effects, butorphanol has sedative effects, which are potentiated by benzodiazepines (i.e., midazolam, diazepam). The combined administration of midazolam and butorphanol is recommended in birds for which midazolam alone provides only an insufficient level of sedation or in birds that require deeper sedation for certain clinical procedures (such as radiographic positioning). Butorphanol can be given in combination with midazolam, drawn into a single syringe, and can be given parenterally as well as intranasally. No side effects of intranasal administration of butorphanol at a dose range of 1–3 mg/kg are seen in psittacine birds.[8] At the University of Wisconsin, we routinely use butorphanol (1–2 mg/kg) combined with midazolam (2 mg/kg) in pet birds administered intranasally or by intramuscular injection. In cockatiels, intranasal sedation with butorphanol (3 mg/kg) combined with midazolam (3 mg/kg) administered intranasally was found to be safe and effective to induce sedation in this species.[10]

Reversal of Sedation

Reversal of sedation in pet birds will depend on the patient and the purpose of sedation. Sedation performed in order to facilitate physical examination and diagnostic sample collection should always be reversed, in order to have the patient return to normal behavior and food intake as soon as possible. It is important not to discharge sedated patients, as owners do not tend to appreciate having a partially sedated bird that might be imbalanced, sleepy, and refusing to eat.

Birds that underwent sedation for e-collar or bandage placement or that were sedated for control of seizures should not be reversed. In these cases, birds should be carefully monitored, and reversal considered if the level of sedation is perceived too deep or the duration of sedation is prolonged and might interfere with physiological behavior, particularly food intake.

Flumazenil

Flumazenil is a benzodiazepine antagonist and is used to reverse the sedative effects of midazolam and diazepam in birds.[2,3,8,9] The injectable form of flumazenil (flumazenil hydrochloride, 0.1 mg/ml; Abaxis Pharmaceutical Products, Schaumburg, IL) can be administered intranasally, intramuscularly, or intravenously without side effects. The recommended dosages range from 0.01–0.1 mg/kg. Alternatively, a flumazenil to midazolam ratio of 13:1 has been recommended, but it has been shown that lower doses of flumazenil achieve complete recovery from midazolam-induced sedation in birds.[2] The author prefers to administer 0.05 mg/kg initially. If reversal is deemed unsatisfactory, then the same amount of flumazenil can be administered repeatedly. Recovery from sedation is usually complete within 10–15 minutes. Reversal with flumazenil is recommended in all birds that are intended to be discharged from the hospital soon after, to avoid injuries from falling off perches and to avoid negative client perception of having a behaviorally abnormal-acting bird. Therefore, it is important to ensure that complete reversal has occurred prior to discharge from the hospital, by assessing complete recovery of the sensory and motor function (e.g., approaching the bird, rotating finger/perch). However, because the half-life of flumazenil is shorter than that of midazolam, re-sedation can occur, and this should be communicated with clients prior to discharge.

Step-by-Step Procedure

1. Educate the bird owner about what to expect in regard to sedation of their bird. Benzodiazepine-based sedation protocols at the published dosages are safe in healthy pet birds, but birds will appear sleepy, which owners are not used to. Inform them that in some cases regurgitation might occur, particularly in macaws.

2. Manually restrain the bird using a towel and immobilize the head. Some owners may be able to administer the sedative drugs intranasally, without requiring manual restraint, which will be less stressful to the bird.

3. Administer the sedative drugs into the nostrils over 5–10 seconds. Split the total dose between both nostrils. If intramuscular administration is performed, administer the sedative drugs by injection in the pectoral muscle using a 25–28-G needle.

4. Release the bird from manual restraint and allow 7–10 minutes for reaching maximum effect of sedation. Do not allow the bird to perch high; instead, place it on the floor or in its carrier.

5. Manually restrain the sedated bird. Monitor respiration and heart rate throughout the restraint period. Perform necessary clinical procedures (e.g., physical examination, blood collection, radiographs).

6. If reversal is indicated, administer flumazenil IM or IN. Split total volume between nostrils. Alternatively, flumazenil can be administered by intramuscular or intravenous injection.

7. Release the bird on the floor or in its carrier and monitor recovery. Do not allow perching high, so as to avoid trauma from falling, until the bird has completely recovered. Most birds will recover within 10–15 minutes. Assess recovery before discharging the bird. Administer a second dose of flumazenil if recovery is incomplete.

Useful Tips

1. Each bird requires an individual assessment prior to choosing a suitable sedative drug protocol. Birds that have had no previous experiences with manual restraint may require less sedation than birds that have had previous negative experiences and thereby become readily stressed or fearful in a veterinary clinic environment.

2. Macaws usually require a combination of midazolam and butorphanol in order to achieve a sufficiently deep level of sedation, suitable for performing a variety of clinical procedures.

3. Even though a bird might not appear sufficiently deep-sedated while under manual restraint for a clinical procedure, it might become more sedated once stimulation is discontinued, making it unsuitable to be discharged from the hospital. Therefore, every bird that is sedated for a procedure and is intended to be discharged shortly after completion of that procedure, should receive flumazenil. Be aware that re-sedation can occur due to the shorter half-life of flumazenil.

4. Cockatoos often do not recover completely after the initial dose of flumazenil. A second dose of flumazenil is often necessary. Requiring a second dose of reversal or re-sedation is less common in other species, but it can occur.

Conclusions

Sedation of pet birds offers a practical alternative to manual restraint of conscious birds and to general anesthesia for a large variety of clinical procedures. Sedation decreases the risks of complications by attenuating the side effects of manual restraint and avoids the potentially fatal side effects of general anesthesia (e.g., cardiovascular and respiratory depression, aspiration of crop content). The intranasal route of administration is an alternative noninvasive technique, which has been proven highly effective in birds and usually increases the client compliance on agreeing to sedation for their bird. Intranasal or intramuscular sedation using midazolam, with or without butorphanol, is a simple, safe, effective, reversible, and readily available option for sedation of avian patients in daily clinical practice.

References

1. Greenacre CB, Lusby AL. Physiologic responses of Amazon parrots (*Amazona* species) to manual restraint. *J Avian Med Surg.* 2004;18:19–22.

2. Mans C, Guzman DSM, Lahner LL, *et al.* Sedation and physiologic response to manual restraint after intranasal administration of midazolam in Hispaniolan Amazon parrots (*Amazona ventralis*). *J Avian Med Surg.* 2012;26:130–139.

3. Vesal N, Eskandari MH. Sedative effects of midazolam and xylazine with or without ketamine and detomidine alone following intranasal administration in ring-necked parakeets. *J Am Vet Med Assoc.* 2006;228:383–388.

4. Sadegh AB. Comparison of intranasal administration of xylazine, diazepam, and midazolam in budgerigars (*Melopsittacus undulatus*): clinical evaluation. *J Zoo Wildl Med.* 2013;44:241–244.

5. Vesal N, Zare P. Clinical evaluation of intranasal benzodiazepines, alpha-agonists and their antagonists in canaries. *Vet Anaesth Analg.* 2006;33:143–148.

6. Gilbert DB, Patterson TA, Rose SP. Midazolam induces amnesia in a simple, one-trial, maze-learning task in young chicks. *Pharmacol Biochem Behav.* 1989;34:439–442.

7. Farkas L, Crowe SF. The role of the benzodiazepine-GABA system in the memory processes of the day-old chick. *Pharmacol Biochem Behav.* 2000;65:223–231.

8. Mans C, Braun J, Feeny M. Intranasal sedation with midazolam or midazolam-butorphanol in psittacine birds: a retrospective evaluation of 114 cases (2010–2012). In: Proc 1st Intl Conf Avian, Herpetol Exot Mammal Med. 2013:292.

9. Prather JF. Rapid and reliable sedation induced by diazepam and antagonized by flumazenil in zebra finches (*Taeniopygia guttata*). *J Avian Med Surg.* 2012;26:76–84.

10. Doss GA, Fink D, Mans C. Assessment of sedation after intranasal administration of midazolam and midazolam-butorphanol in cockatiels. *Am J Vet Res.* 2018;79:1246–1252.

SPEAKER INFORMATION
(click the speaker's name to view other papers and abstracts submitted by this speaker)

Grayson A. Doss, DVM, DACZM
School of Veterinary Medicine
University of Wisconsin-Madison
Madison, WI, USA

Christoph Mans, DrMedVet, DACZM, DECZM (ZHM)
School of Veterinary Medicine
University of Wisconsin-Madison
Madison, WI, USA

URL: https://www.vin.com/doc/?id=9768898

Getting Them Back on Their Feet—Identification, Causes and Treatment of Pododermatitis

ExoticsCon Virtual 2020 Proceedings

Ellen Rasidi, BBiomedSc, BSc (Vet) (Hons), BVSc

Department of Conservation, Research and Veterinary Services, Wildlife Reserves Singapore, Singapore

Abstract

Pododermatitis is considered a prevalent disease of captivity and is generally not found in wild birds. While most commonly described in captive raptors, pododermatitis can occur in any species of captive bird. This masterclass will cover the following topics, which will be illustrated using case studies from Jurong Bird Park, one of the largest bird parks in the world, with a collection of nearly 3500 birds across 400 species.

Classification of pododermatitis lesions include ulcerative, granulomatous, hyperkeratotic, fissural, nodular, papillomatous and differing degrees of severity.

Environmental and husbandry-related causes can include housing, climate, perches and substrate, diet and nutrition, activity. Development of pododermatitis can also be secondary to primary traumatic or medical conditions, or as a result of medical or surgical interventions, which must be considered when there is any defect of the appendicular skeleton.

There are differences in the presentation of acute versus chronic pododermatitis, and potential complications and consequences of both forms. Acute presentations are often associated with infectious repercussions, and chronic presentations are associated with inflammatory sequelae, such as amyloidosis. Differences in the management of acute versus chronic pododermatitis also include radical versus conservative treatment.

Treatment options include conventional surgical and medical interventions, as well as complementary procedures such as nerve blocks, intralesional and regional antibiotic therapy, and photodynamic therapy and there are advantages and disadvantages of each of these approaches.

Biomechanical treatment options include substrates and perching options, and pressure-relieving bandaging. Bandaging includes conventional bandage styles, donut-style disposable shoes, foam plantar shoes, padded reusable textile shoes, and molded silicone casting. A novel 3D printed option for silicone shoes will also be described.

Introduction

Pododermatitis, also referred to as foot pad dermatitis, or by the lay term bumblefoot, is a significant disease syndrome of the avian foot and one of the major diseases affecting the locomotory system. It is generally a chronic inflammatory or degenerative condition initiated by underlying abnormalities of pressure or trauma on the plantar surface of the foot, usually the metatarsal or digital pads. This results in plantar lesions associated with the development and progression of pressure sores; tissue swelling, calluses, abrasions, ulcerations, and fissures.

Hyperkeratosis, abrasion and avascular necrosis arising from initiating factors, if unnoticed or treated inappropriately, may progress to vascular damage, callus formation, inflammation, cellulitis, ulceration, and bacterial colonization of the skin and subcutaneous tissues of the plantar surfaces of the feet.

If left untreated, the disease becomes progressive, invasive, and eventually disabling, and chronic cases may develop deep bacterial infections, abscesses, osteomyelitis, synovitis, bacteremia and fatal septicemia. In non-fatal cases, lameness and loss of limb function may be caused by local infection or by osteotic changes attributed to persistent abnormalities in forces affecting the foot.

Pododermatitis can have significant welfare and implications for affected individuals and populations. While cases may range from mild, barely visible skin changes, to severely debilitated birds, most clinical signs of pododermatitis are considered painful or at least causing discomfort. Early stage hyperkeratotic lesions may result in mild lameness but may develop into more severe lesions which are associated with visible signs of pain and concurrent behavioural changes. Birds suffering from pododermatitis may be reluctant to stand, perch on certain surfaces, walk as much as usual, or may display other behaviours associated with pain, such as weight shifting, hock-sitting, recumbency, or displacement behaviours such as over preening or self trauma. Normal daily maintenance activities may also be adversely affected, with caregivers reporting reductions in self-preening, social engagement, and food and water consumption.

In commercial poultry production systems, the presence of pododermatitis lesions is used as an indicator of flock welfare throughout the industry and is important commercial consideration. Affected birds show reduced weight gain, and lesions can act as a portal of pathogen entry, with implications for food safety and carcass contamination. Pododermatitis also has negative economic impacts on other avian industries, including falconry, zoological collections and educational presentation programs.

As the etiology of pododermatitis generally involves abnormal bird-environment interactions, pododermatitis is usually considered a disease of captivity and is rarely found in wild birds, unless as a sequela to injury. There are very few reports of pododermatitis in free ranging wild birds, in contrast to their captive conspecifics.

Pododermatitis is also often reported in wild birds in temporary captivity undergoing rehabilitation. Along with fungal respiratory disease, it is one of the most frequent and important clinical complications in rehabilitation centers, particularly in seabirds. It is less frequently described in pet birds, but may be seen particularly in captive-reared, older, overweight pets, or those kept in suboptimal conditions. As in other groups, it is also reported as a consequence of other disorders affecting the pressure placed on the feet. Many early reports of pododermatitis come from captive birds of prey, and the disease remains commonly reported in falconry birds, those undergoing rehabilitation and in zoological collections. It is almost universally described in avian production systems (with a prevalence of between 20–98%), regardless of the species farmed, but more particularly in broiler breeds of chickens and ducks, and turkeys—where the condition has a direct effect on carcass quality—rather than in layer breeds.

Pododermatitis can become a significant issue in a large variety of avian taxa in large zoological collections, where shared environments can result in whole flocks being affected. There are several examples in the literature describing pododermatitis issues in populations of penguins, flamingos and other waterbirds.

At Jurong Bird Park, one of the largest bird parks in the world, with a collection of nearly 3500 birds across 400 species, pododermatitis cases have been reported in species representing most orders of birds. Examples and case studies from the Jurong Bird Park Veterinary Department will be used to illustrate the main concepts in this masterclass.

Classification of Pododermatitis Lesions

The classification of pododermatitis lesions is necessary to evaluate the type and severity of the condition and is used for many different reasons.

In production systems, lesion classification is used as part of carcass scoring and evaluation, and also to evaluate welfare. In wildlife rehabilitation, classification is used to evaluate both the rehabilitation process, techniques and protocols, and also as a prognostic indicator or successful rehabilitation and release. In collection and pet birds, lesion classification is used not only to evaluate enclosure suitability and substrate, but also to determine appropriate treatment and management techniques.

Multiple classification systems have been described and are in common use, but generally fall into two types; those that describe the severity of the lesion and those which describe the type of lesion. Some classification systems encompass both severity and lesion characteristic. However, both types of classification systems describe macroscopic pododermatitis lesions, and therefore a limitation of visual classification systems is that the condition is only detected once the lesions have formed and disease is well established, and generally does not distinguish between active and inactive (healed or fibrosed) lesions, nor do the classification systems take into account subclinical disease.

Identification of subclinical plantar disease would be beneficial, to allow early initiation of husbandry changes and treatment to prevent the development of clinical lesions. This has been attempted with thermal imaging and infrared thermography in flamingos and penguins but has not been shown to be a reliable diagnostic tool.

Classification systems describing the severity of lesions are generally based on the classic pododermatitis lesion. This is a nodular lesion with central ulceration, also known as the ulcerative form of pododermatitis. The first system, reported in the commercial poultry industry in 1984, is the most simplistic and assigns a score from 0 to 3, according to the following descriptions:

Score	Lesion
0	None
1	Small scab(s) <5% pad area
2	Larger scabs <25% pad area
3	Severe, large scab-filled ulcers

Pododermatitis lesions in poultry generally begin as plantar skin discolouration, developing to small scaly brown scabs on the metatarsal and digital pads (Score 1). These lesions develop and enlarge (Score 2) through mechanical erosion and inflammation, leading to hyperplasia and epidermal necrosis (Score 3). Necrotic debris is often associated with foreign material from the environment or substrate, and bacteria.

In captive collection or pet birds, classification systems are generally highly descriptive of the plantar lesions, as they are more often used for clinical assessment, prescription and evaluation of treatment. The systems used are based on those originally described for the description of pododermatitis in captive raptors, and also range in the number of categories of severity described.

A review article in JAMS by HF Burke *et al.* in 2002 compares four similar classification systems for pododermatitis in raptors, ranging in complexity from 3 to 7 stages of lesion severity. These systems are reproduced here:

Score	Halliwell WH (1975). Bumblefoot infections in birds of prey. *J Zoo Anim Med.* 1975;4:8–10.	Redig PT. Bumblefoot treatment in raptors. In: Fowler ME, ed. *Zoo and Wild Animal Medicine.* 3rd ed. Philadelphia, PA: WB Saunders; 1993:181–188.	Remple JD. Raptor bumblefoot: a new treatment technique. In: Redig PT, Cooper JE, Remple JD, Hunter DB, eds. *Raptor Biomedicine.* Minneapolis, MN: University of Minnesota Press; 1993:174–179.	Degernes LA. Trauma medicine. In: Ritchie BW, Harrison GJ, Harrison LR, eds. *Avian Medicine: Principles and Application.* Lake Worth, FL: Wingers; 1994:417–433.
1	Serious chronic infection, cellulitis and swelling of the entire metatarsal pad, always chronic, requires aggressive surgical and medical treatment, and usually carries a guarded to poor prognosis.	Grossly enlarged metatarsal pad with infection and cellulitis of the entire metatarsal pad, the most severe form, it is always chronic and yields a guarded prognosis, includes trap-induced injuries.	Early insult or lesion of a prominent plantar area with no apparent underlying infection: a bruise, early ischemia, or hyperkeratotic reaction (corn, callus).	Desquamation of small areas of the plantar foot surfaces represented clinically by the appearance of small, shiny pink areas—peeling or flaking of the skin on the legs and feet.
2	More localized, encapsulated lesion of the metatarsal pad.	Enlargement of the metatarsal pad with a localized encapsulated central lesion, prognosis is fair after drainage, irrigation, and closure of the wound when infection is no longer apparent.	Infection of underlying tissues in direct contact with the surface lesion but no gross swelling: puncture (with localized infection), local ischemic necrosis (a penetrating corn or scab).	Smooth, thinly surfaced, circumscribed areas on the plantar metatarsal pads of one or both feet with the subcutaneous tissue almost visible through the translucent skin; no distinct ulcers are recognized.
3	Discrete lesion of the foot, such as foreign body reaction, small cut or puncture wound, or excessive wearing or build-up of epithelium on the plantar surface of the foot.	Slight enlargement and inflammation of one discrete area of the foot, either the metatarsal pad or digital pads.	Infection with gross pedal inflammatory swelling—serous, fibrotic, or caseous.	Ulceration of the plantar metatarsal pads, in some birds, a peripheral callus may form.
4		Enlargement of a distal portion of one of the toes without rupture of the epidermis, from sitting on an improperly sized perch, also associated with rupture of the flexor tendons, prognosis is good to fair with early treatment.	Infection with swelling of underlying tissues involving deep vital structures; chronic, producing tenosynovitis, arthritis and/or osteomyelitis, when synovial structures are involved, tissue exudate often combine with purulent synovium: fibrotic or caseous.	Necrotic plug of tissue present in ulcer, most species with ulcers and accumulation of necrotic debris exhibit pain or mild lameness.
5			Crippling deformity and loss of function.	Swelling and edema (cellulitis) of the tissues surrounding the necrotic debris, the digits or foot may also be edematous, necrotic debris may start to accumulate in the metatarsal area, suggesting infection of the tendon sheaths, severe lameness is common, the

				entire metatarsal pad may be affected, this is generally a chronic lesion.
6				Necrotic tendons recognized clinically as swelling in the digits and ruptured flexor tendons, ankylosis and nonfunctioning digits usually present in recovery.
7				Osteomyelitis.

While each study, report, clinic or institution may use a different classification system depending on the species examined, type of practice and need, this is unlikely to be problematic as long as the use of the selected classification system is consistent, and well defined. As in the commercial poultry industry, it is recommended that the classification system is well publicized or displayed within the clinic or institution, preferably with pictorial examples.

Classification systems describing the type of lesion are generally applied to flamingos, a species which appear to develop a variety of lesions. This classification system recognizes four pathoanatomical categories: the classic nodular lesion, hyperkeratosis, fissural, and papillomatous lesions. Lesions may also be identified as a combination of two or types, and multiple types may be present in the same individual, particularly in very chronic case. Within this classification system, severity can also be indicated in fairly truncated form; generally, only mild (Score 1) or advanced (Score 2) stages are indicated. As in previously described classification systems, Score 0 indicates the absence of macroscopic lesions.

Score	Nodular	Hyperkeratotic	Fissural	Papillomatous
1	Closed lesion with no central necrotic plug	Epithelium is flattened or slightly overgrown	Fissures less than 2 mm deep	Small, fingerlike proliferations
2	Open lesion with exposed necrotic tissue	Epithelium shows marked overgrowth	Fissures deeper than 2 mm	Clusters or lumps of proliferations

While this classification system is generally only reported in the assessment of pododermatitis in flamingos, it can also be applied to other species in which non-classic or mixed-type lesions may be seen. To increase accuracy in lesion description, the plantar surface of each foot can also be divided into numbered weightbearing regions. In flamingos, there are seven such regions; the central metatarsal pad and two digital pads on each digit, one proximal and one distal. The number and pattern of these weightbearing regions varies based on the pedal anatomy of each species.

Figure 1

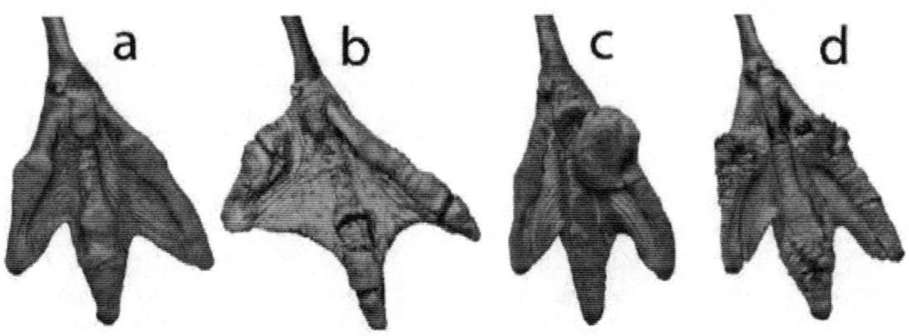

Classification of foot lesions in captive flamingos: a) hyperkeratosis, b) fissures, c) nodular lesions, and d) papillomatous growths. From: Nielsen AMW, Nielsen SS, King CE and Bertelsen MF (2010). Classification and prevalence of foot lesions in captive flamingos (Phoenicopteridae). *J Zoo Wildlife Med.* 41(1):44–49.

Causes of Pododermatitis

Environmental and Husbandry-Related Causes

As primarily a disease of captivity, the most common etiology of pododermatitis is directly due to the environment in which the bird is kept. Pododermatitis lesions are pressure sores caused by abnormalities of pressure or trauma on the plantar surface of the foot, and so the lesions are generally initiated in the regions in direct contact with the perches or substrate—the metatarsal or digital pads.

Like many aspects of avian anatomy, the pelvic limb has evolved to maximize efficiency in a particular environmental and behavioural niche. Basic examples of this include the placement of the pelvic limb on the body, which is dependent on whether the bird spends more time perching or standing and weight-bearing as opposed to swimming or flying (diving ducks have legs placed more caudally than dabbling ducks); the enormous variation in pedal morphology (aniso-, zygo-, hetero-, syn- and pamprodactyly, as well as variations in palmation and lobation); and the species variation in the number of digital actuator muscles (extensors and flexors), ranging from 11 to 18. Because of this variation and adaptation for particular niches, it is inevitable that there will be limitations in the extent to which the captive environment can permit normal free-ranging behavior. Even in the highest ideals of enclosure design, the very act of limiting the activity of the bird to a restricted space will lead to deviations in the normal forces acting on the feet. Whether or not these deviations are sufficient over time to cause clinical pododermatitis depends on the extent and chronicity of the imposed pressure abnormalities, as well as other confounding environmental factors.

The most obvious causes of pododermatitis lesions are the surfaces on which the bird stands or perches (weight-bearing substrates), and there are three main characteristics of substrates which may predispose to abnormal plantar pressures. The first is hardness. Hard, solid surfaces such as concrete, rock and tiles do not compress with the weight of the bird and therefore result in the highest pressure returned to the plantar surface. Natural surfaces such as wood produce lower pressures, and substrates such as earth and turf compress easily and therefore reduce the pressure on the foot, as do artificial compressible substrates such as foam and rubber matting. The second characteristic of substrates which can predispose to pododermatitis lesions is texture. A bird weight-bearing on a smooth solid substrate will experience pressure at the same points on the plantar surface regardless of where they stand; in pet birds this is most often seen where commercial dowel perches are used. A textured surface like a gnarly natural branch will change the pressure points on the plantar surface with each step, thus distributing the pressure on the foot over time. The third aspect of a substrate that impacts plantar pressure is the movement of the substrate itself. Movement of generally encourages the bird to change its stance and shift its weightbearing, thus actively changing the pressure points on the foot. Example of moveable substrates are dry sand and gravel, rope perches and dynamic branch perches. In summary, the harder, smoother and more rigid the substrate, the more likely it is to contribute to the formation of pressure sores and pododermatitis lesions.

However, there are a multitude of factors which also influence the likelihood of pressure sores and progression of pododermatitis. Being overweight or obese has been shown to increase the incidence or severity of pododermatitis lesions, likely simply by increasing the pressure borne by the feet in normal weightbearing. Excess bodyweight or over conditioning may also be associated with other risk factors typically found in captive birds. The sedentary lifestyle of captive birds when compared to their wild counterparts may firstly predispose the captive bird to overconditioning or obesity through reduced energy expenditure or increased caloric intake (or often a combination of both). In addition, the sedentary captive lifestyle results in increased time perching or standing, particularly in the same position or location, thus increasing the likelihood of plantar pressure sore formation. Reduced energy expenditure, increased caloric intake and limited perching positions may also result from housing in enclosures of insufficient size to allow appropriate activity. Enclosures may not only have perching substrates of inappropriate material and texture, but insufficient perching space or options. Food may be readily available without effort made on the bird's part. Think of the stereotypical Victorian-era parrot in a cage; the food bowl is directly in front of a sedentary bird who has a single perch and no room to open its wings, let alone take more than two steps in either direction. While this is the extreme situation, there are elements of similar issues in many modern aviaries and enclosures.

As well as food availability, diet composition may also increase the risk of pododermatitis lesion development. While overall increased caloric and macronutrient intake lead to increased weight, deficiencies in other nutrients may reduce skin integrity, thus predisposing to lesion formation. In commercial poultry flocks, higher prevalence of pododermatitis has been reported in turkeys fed a diet deficient in methionine, and also in those with diets deficient in biotin. Zinc deficiency has also been hypothesized as having a role in the development of pododermatitis in flamingos, and it is possible that deficiencies in other vitamins and amino acids involved in skin formation and maintenance (such as vitamins A and E, pantothenic acid, lysine and riboflavin) may also play a role.

Another less obvious effect of diet on predisposition to pododermatitis is water and indigestible fat content, and the contribution to faecal composition. This is generally only applicable to individuals and species housed in enclosures where the substrate quality is directly affected by the composition of the droppings. The most common example is commercial poultry housed in deep litter systems, but might also apply to backyard or pet poultry, or handraised birds kept on organic litter. In commercial poultry operations, litter wetness is increased where birds are fed diets with higher indigestible fats. The fats passed in the faeces also causes both the faeces and litter to stick to the feet, increasing skin damage and increasing ammonia levels. The design and placement of water bowls and drinkers also directly impacts litter wetness through spillage. Wet litter not only leads to maceration and degradation of the plantar skin but increases the risk of infection of foot lesions. Pododermatitis may be a sequela of moist plantar dermatitis (contact dermatitis) and in large commercial poultry flocks, often most strongly associated with drinker type, litter depth and litter material.

There have also been some suggestions that climate may be a predisposing factor in the development of pododermatitis in certain captive species, though this is so far inconsistent. Humidity may increase in the incidence of moist plantar dermatitis in some species, though in a study of captive flamingos (a wetland species), temperature and humidity, as well as indoor versus outdoor housing, were found to correlate with the development of certain types of pododermatitis lesions. Conversely, an environment that is too dry may also predispose to the development of pododermatitis by causing desiccation of the plantar skin.

Secondary Pododermatitis

In addition to environmental and husbandry-related causes, pododermatitis may also develop secondary to any primary traumatic (including thermal and electrical burns, and frostbite) or medical conditions, or as a result of medical or surgical interventions in which plantar pressure is altered.

In the rare cases of pododermatitis reported in wild free-ranging birds, most appear to be unilateral cases which have developed due to injury of the contralateral pelvic limb, causing increased use and therefore pressure on the affected foot. A similar etiology occurs in captive birds, and pododermatitis lesion development has been reported in numerous cases after orthopaedic injury or surgery, as a consequence of asymmetric weight-bearing during recovery. Amputation of any part of the pelvic limb results in risk of pododermatitis occurrence on the contralateral foot.

Alterations in weight-bearing and distribution, and resultant pododermatitis, may also be caused by anatomical aberrations beyond the pelvic limb, in either the axial or appendicular skeleton or musculature. Lateral scoliosis of the spine may also lead to unilateral plantar pressure sores, as may any other asymmetry of the trunk that affects pelvic placement. Asymmetry of the wings or wing function may also cause mild alterations in weight-bearing, and lead to the development of pododermatitis over time. In the author's observation, pinioning does not appear to increase the incidence of pododermatitis, though more severe asymmetries pectoral limb function do, such as amputation at the humerus, or ankylosis of the elbow, preventing limb extension. Both these abnormalities prevent the wing from being used for balance, and it is suspected that it is this that causes the asymmetry in weight bearing.

Pododermatitis may also be initiated or complicated by infection or inflammation to the foot. Moist dermatitis is mentioned previously, where infection is seeded via devitalized macerated plantar skin, though infection of the plantar skin and underlying structures may also occur through devitalization from bruising and pressure necrosis, or direct inoculation through skin puncture. The most common causes, particular in raptors and poultry, include progression of avipoxvirus lesions, and self-inflicted penetrating injuries from overgrown talons.

Though pododermatitis is not an infectious disease, a variety of viral, bacterial and fungal pathogens have been isolated from pododermatitis lesions in a range of species. Virus-like particles suggestive of Papovaviridae have been noted in a northern gannet with proliferative pododermatitis, and in multiple waterfowl species with verrucous hyperkeratotic lesions; herpesvirus-related proliferative lesions have been reported in macaws and cockatoos; herpesvirus has also been associated with ulcerative foot lesions in a duck. Mycotic (*Candida* spp.) pododermatitis is reported in turkey poults. Aerobic bacterial agents are commonly isolated from pododermatitis lesions and may play more of a causative role in younger birds. The most common bacterial isolate is the ubiquitous *Staphylococcus aureus*, often identified with *Escherichia coli*, *Corynebacterium* species, *Enterococcus faecalis*, *Pseudomonas* and *Proteus* species. Less commonly reported are *Aerococcus viridans*, Group D *Enterococcus*; *Bacillus*, *Moraxella*, *Pasteurella*, *Klebsiella*, *Clostridium*, *Diplococcus*, *Nocardia*, *Actinobacillus*, *Actinomyces*, *Aeromonas* and other *Staphylococcus* and *Streptococcus* species.

Bacterial invasion of the stratum corneum and exocytosis of heterophils into the epidermis are noted to be an early histologic signs of pododermatitis. In later, clinically visible stages, pressure necrosis and reduced vascular perfusion reduce the efficacy of local immune response, as well as antibiotic delivery to the affected area. A generalized failure to degrade phagocytosed pathogens then leads to chronic granulomatous disease.

Acute versus Chronic Pododermatitis

Presentation

Pododermatitis is considered a chronic disease, though cases may present with acute complications, and therefore require alterations in treatment and management strategies. In most instances, the authors consider any case presenting with recent trauma, active bleeding or active infection as an acute case, requiring immediate medical and potentially surgical management. Chronic cases, where infection is minimal and localized, and the lesions are granulomatous rather than purulent, can be managed through relief of pressure and through environmental changes.

Different presentations will also differ in disease considerations. For example, in acute cases, the immediate concerns are usually related to infection; local spread to pedal structures, osteomyelitis, systemic spread and bacteremia/septicemia. In chronic cases, concerns may be more centered around the consequences of ongoing and persistent inflammation, particularly amyloidosis, or other effects of chronic disease, such as immunosuppression and anemia.

However, as pododermatitis cases may present at any time on the disease timeline, they do not necessarily fall into easy categorizations of 'acute' or 'chronic' but are likely a combination of both. In individual cases, the bird should undergo thorough clinical examination and assessment to ascertain the general health and condition of the bird, as well as to identify the extent of any infection or inflammation. Earlier-stage or more acute cases may show inflammatory leukograms with elevations in CK; chronic cases may show mild anaemia and monocytosis. Pedal radiographs are always advised to ascertain bony involvement. Lesions should then be characterized and classified using whatever classification system is used within the particular clinic or institution. Culture of any open lesions is generally unrewarding; culture of deep samples, such as synovial fluid aspirates, may be more useful, though penetration of the plantar region by systemic antibiotics is limited, and are most effective in cases of osteomyelitis, or systemic infection. Ultrasonography or advanced imaging may be used to assess tendon integrity if required. It should be noted that though *Clostridium tetani* toxicosis is rarely reported in birds, it was been reported in gyrfalcon associated with chronic pododermatitis lesions.

Management

Apart from direct medical and surgical treatments, it is imperative that a thorough examination and evaluation be performed of the bird's captive environment, to establish the (potentially numerous!) contributing factors. It is important to remember that while the clinical disease may be able to be resolved in the individual, unless the causative environmental factors are addressed, the condition will recur.

Treatment Options

As pododermatitis is a complex disease with a multifarious etiologies and contributing factors, the treatment of pododermatitis may require multiple approaches over a long period of time. No one treatment modality is likely to resolve the issue on its own. However, the main concept of pododermatitis treatment revolves around the relief of plantar pressure and allowing gradual healing of open wounds and lesions. Generally speaking, the more chronic the lesion, the longer the treatment time; in very chronic and severe cases, this may be up to a year. While historical literature on pododermatitis treatment tends to emphasis surgical exploration, debridement and primary closure, recent trends in treatment espouse healing by secondary intention, as tissue tension and recurrence rates are lower.

In addition to medical and surgical interventions, underlying causes must be addressed and rectified to avoid recurrence.

Medical Interventions

Medical intervention is almost always required in cases with acute complications: active infection or inflammation, tendon and bony involvement, purulent or caseous discharge or abscessation. Not all abscesses are associated with infection; one study of pododermatitis abscesses in penguins showed that around 10% were sterile.

Pododermatitis is always assumed to be a condition that causes pain, or at least discomfort. Depending on the species and the individual health of the bird, appropriate analgesics may range from non-steroidal anti-inflammatories to opioids, depending on the extent of the lesions, the involvement of pedal architecture and the behavior and response of the bird. Pain management is particularly important in unilateral pododermatitis, where pain in the affected limb may cause overreliance on the contralateral limb for weightbearing, thus leading to the development of pododermatitis lesions on the previously unaffected foot.

If surgical debridement or intervention is required, local and regional nerve blocks should be considered during the intra- and perioperative period, as nociceptors in tendons and joints (deep pain receptors) are often stimulated.

Antibiosis of some modality is often indicated, and may be a combination of topical (intralesional), systemic or regional through intravenous regional limb perfusion (IVRLP) or intraosseous regional limb perfusion (IORLP), depending on the type and severity of lesion, other treatment used, and patient (or owner) compliance or cooperation. Topical or intralesional antibiotics or antimicrobials are most often used after debridement and may also be used in conjunction with bandaging treatments following such a procedure. Antibiotic gels, creams and ointments are all described in the treatment of open lesions, and those with enzymatic debridement properties may also be useful, depending on the extent of tissue damage or penetration. Hydrocolloid gels are also useful in cases where there is significant tissue exposure and desiccation of underlying structures is on concern. Silver sulfadiazine cream is effective against a wide range of organisms and also promotes re-epithelialization, and so is more appropriate for more superficial lesions. Manuka honey has also been used with success, in conjunction with appropriate bandaging.

There are some reports of the use of intralesional antibiotic-impregnated polymethyl methacrylate (PMMA) beads after debridement, and antibiotic-impregnated poloxamer gels may also be used. Both these methods achieve ongoing high concentrations of antibiotic in a region where vasculature perfusion may be compromised.

Antibiotic therapy is best guided by culture of deep lesion swabs, and sensitivity testing. Pododermatitis lesions are often highly contaminated wounds and testing most often reveals environmental pathogens or organisms associated with faecal contamination. Systemic antibiotic therapy should be considered to prevent spread of local infection and ascending or systemic infection which may lead to sepsis. Broad spectrum antibiotics may be used initially, then changed according to results of culture and sensitivity testing as appropriate. Systemic antibiotic therapy may be continued for several weeks, depending on the extent of wound contamination. Topical treatments are often continued longer.

Antimicrobial photodynamic therapy (PDT) has been suggested as an alternative therapy to conventional antimicrobial treatments for pododermatitis. PDT is applied directly to the infected area, selectively destroying a large number of microorganisms without causing tissue destruction and may be useful in birds where the risk of wound contamination is high, such as penguins. It also has significant potential in infections with resistant microbes.

Low level laser therapy may be used as an adjunct treatment to assist in wound healing and pain relief. It is used in the treatment of human diabetic foot ulcers and is widely used in veterinary medicine in some places.

Surgical Interventions

Surgical treatment of pododermatitis is based around surgical debridement of the lesion and management of the open wound. Surgical closure is rarely recommended due to the scarcity of available skin for closure, significant tension and risk of dehiscence, and higher recurrence rates. Primary closure may not be possible due to the size of the lesion(s), and the use of purse string or tension-relieving sutures should only be used if absolutely necessary. Repeated attempts may result in exuberant granulation tissue or scar tissue formation at the lesion site. Use of skin flaps utilizing interdigital webbing is reported.

Healing by secondary intention, where there is no damage to the underlying bone and tendon structure of the foot, maintains the normal architecture of the foot and range of motion of the joints, and is generally preferred if possible. Cases in which there is osteomyelitis or damage to the bones, tendons or ligaments tend to have a poor prognostic outcome.

Debridement is performed under general anaesthesia. Local and regional nerve blocks are often useful at this stage. The foot should be cleaned of all debris and aseptically prepped as best as possible to avoid intraoperative contamination of deep tissues. All caseous purulent material should be removed, as well as all devitalized tissue and foreign debris. Flushing and irrigation are recommended to reduce antigen load. All wound edges should be freshened to encourage wound healing.

The wound healing process is a complex continuum of the inflammatory, proliferation, and remodeling phases of healing, and depends on the selection of appropriate topical antimicrobials or healing stimulants, dressings, and protective bandages.

Bandaging

Bandaging has two primary purposes. In the early stages of healing, the bandage must protect the open plantar wound after surgical debridement and should be changed regularly to allow monitoring of healing, cleaning or further debridement of the wound, and reapplication of topical treatments. The bandage or dressing must also alleviate direct pressure on lesions and redistribute the pressure on the plantar surface to prevent worsening of an existing lesion or recurrence of a debrided lesion. This becomes even more important in the chronic, remodeling phase of wound healing, which may take several months, as new tissue growth responds to pressure of weight-bearing.

Low-grade pododermatitis lesions without ulceration or contamination, particularly those involving flattening of the plantar papillae or hyperkeratosis, may respond to pressure-relieving bandaging only, which can be carried out while the underlying environmental causes are examined and rectified.

- Small lesions may be treated with commercially available corn pads, which redirect pressure to the immediate area surrounding the lesion.
- Interdigitating bandages are designed to provide padding to the central tarsal area. They allow relatively normal movement and use of the digits for perching. They are not suitable for birds with interdigital webbing or non-perching birds.
- Ball bandages are an extreme form of padding where all digits are closed around a central padded sphere. This provides padding to the central tarsal area and transfers pressure partially to the dorsal aspect of the digits, depending on the size of the ball.
- Donut bandages maintain the digits in full extension and relieve pressure on the tarsal area, by transferring weight-bearing to the digits. The donut may be made of rolled

bandage material, but most commonly foam, which can be cut to size depending on the length of the digits and the size of the central lesion. A common method is to use slices of 'pool noodle' flotation devices, as these have an absent central core and are an ideal diameter for many hawks and falcons. Donut bandages are ideal for tarsal lesions, but not for digital lesions, and as weigh bearing pressure is transferred to the digits, care must be taken to ensure that secondary pressure sore do not develop.

- Foam shoes are comprised of a thick foam sole out of which holes are cut to accommodate the placement, size and shape of the lesion(s). These are therefore suitable for birds with webbed feet or those that usually stand on flat ground.
- 'Íceflo' shoes are a version of the foam shoe, using thick neoprene material instead of the foam sole, and a covering bootie—adapted from a wetsuit boot and fastened with Velcro—instead of the covering bandages. This style shoe was designed for penguins, as it can be worn both while swimming and on land. It is not suitable for other aquatic birds like Anseriformes, as paddling motions tend to dislodge the shoes, and the retention of moisture on land causes maceration and moist dermatitis of more delicate webbing.
- Silicone foot casts have been described for post-operative treatment and are designed to provide a soft but semirigid supportive cushion without creating pressure ischemia, by evenly distributing pressure over the whole surface.
- 3D-printed silicone shoes have been tested by the authors recently, based on the concept of the donut and foam shoe styles of bandaging. The silicone shoe is designed to be customizable, reusable, and comfortable, without the use of bandaging material. The benefit of this style is less wastage of disposable bandaging material, and the ability to get the shoe wet without the bandaging requiring replacement. Cutouts in the sole are created to accommodate the lesions.

Husbandry Considerations

Other pressure-relieving treatment options to consider, as well as bandaging, regard the substrate and perching options offered to the bird. Substrate modification may often be part of a conservative treatment option for low-grade pododermatitis lesions, or part of a larger environmental modification to rectify causes of lesion development.

As well as more permanent modifications to perches described earlier, during convalescence perches may be padded with bandage material, thin foam (such as yoga mats), rope, artificial turf, indoor-outdoor carpeting, neoprene or other soft material. For birds requiring extensive or prolonged bandaging, the replacement of perches with platforms may be necessary. Caged birds that use the cage mesh for climbing may benefit from a cage that uses a majority of vertically oriented bars, as there is far lesser plantar pressure when these are gripped.

For birds who do not, or are unable to perch, coarse-grain sand is an excellent substrate for recovery and to prevent recurrence. The shifting surface encourages muscle activity in the foot and pelvic limbs, and constantly changing pressure on the plantar surfaces. However, it must be dry, as wet sand quickly becomes hard and smooth, and very fine sand often compacts and hardens as well. For wading birds, sand or mud can also be used as an underwater substrate in the same way.

Rubber matting is also a good option for convalescent aviaries, either as a floor covering or pond lining. Saline foot baths have been used with success in several instances, either as a daily individual antimicrobial treatment, a walk-through as part of a pond system, or as a temporary water change for a period of time. This is an excellent option for treatment of larger flocks or wild birds undergoing rehabilitation, where regular restraint for treatment is challenging or not advised.

Activity is also important in the treatment of and prevention of pododermatitis. In birds that are able to fly or swim, these activities should be encouraged as much as possible, as it not only eliminates weight-bearing, but also increases blood flow. Resolution of lesions is much quicker in penguins that are encouraged to spend more times in the water and in hawks that spend more time in flight training. These birds are also less likely to become overweight, which also increases plantar pressure. For birds unable to do either, encouraging activity with enrichment items is recommended, particularly those which encourage whole body movement.

Conclusion

Pododermatitis is an important medical and welfare concern in captive birds of all species. It is a complex presentation with a multifactorial etiology and has many treatment and preventive challenges. Successful resolution and prevention of recurrence will always require a multimodal approach, as well as commitment to long-term treatment by the caregiver.

References

1. Bailie CL, Baxter M, O'Connell NE. Exploring perch provision options for commercial broiler chickens. *Appl Anim Behav Sci.* 2018;200:114–122.

2. Beaufrère H, Laniesse D, Stickings P, Tierney R, Sesardic T, *et al.* Generalized tetanus in a gyrfalcon (*Falco rusticolus*) with pododermatitis. *Avian Dis.* 2016;60:850–855.

3. Bueno I, Anderson G, Willette M, Redig PT, Ponder J. Distraction osteogenesis in two wild raptors. *J Avian Meg Surg.* 2019;33:427–436.

4. Burke HF, Swaim SF, Amalsadvala T. Review of wound management in raptors. *J Avian Med Surg.* 2002;16:180–191.

5. d'Ovidio D, Noviello E, Adami C. Nerve stimulator-guided sciatic-femoral nerve block in raptors undergoing surgical treatment of pododermatitis. *Vet Anaesth Analg.* 2015;42:449–453.

6. Daoust P-Y, Wadowska D, Kibenge F, Campagnoli RP, Latimer KS, *et al.* Proliferative pododermatitis associated with virus-like particles in a northern gannet. *J Wildlife Dis.* 2000;36:378–382.

7. Doneley RJT, Smith BA, Gibson JS. Use of a vascular access port for antibiotic administration in the treatment of pododermatitis in a chicken. *J Avian Meg Surg.* 2015;29:130–135.

8. Duncan AE, Torgerson-White LL, Allard SM, Schneider T. An evaluation of infrared thermography for detection of bumblefoot (pododermatitis) in penguins. *J Zoo Wildlife Med.* 2016;47:474–485.

9. Erlacher-Reid C, Dunn JL, Camp T, Macha L, Mazzaro L, Tuttle AD. Evaluation of potential variables contributing to the development and duration of plantar lesions in a population of aquarium-maintained African penguins (*Spheniscus demersus*). *Zoo Biol.* 2012;31:291–305.

10. Fiorello CV. Intravenous regional antibiotic perfusion therapy as an adjunctive treatment for digital lesions in seabirds. *J Zoo Wildlife Med.* 2017;48:189–195.

11. Gartrell BD, Collen R, Dowding JE, Gummer H, Hunter S, *et al.* Captive husbandry and veterinary care of northern New Zealand dotterels (*Charadrius obscurus aquilonius*) during the CV Rena oil-spill response. *Wildlife Res.* 2013;40:624–632.

12. Gebhardt-Henrich SG, Toscano MJ, Wurbel H. Perch use by broiler breeders and its implication on health and production. *Poultry Sci.* 2017;96:3539–3549.

13. Jones MP. Vascular diseases in birds of prey. *J Exotic Pet Med.* 2013;22:348–357.

14. Jones MP. Selected infectious diseases of birds of prey. *J Exotic Pet Med.* 2006;15:5–17.

15. Kajagar BM, Godhi AS, Pandit A, Khatri S. Efficacy of low level laser therapy on wound healing in patients with chronic diabetic foot ulcers—a randomised control trial. *Indian J Surg.* 2012;74:359–363.

16. Klambeck L, Stracke J, Spindler B, Klotz D, Wohlsein P, Schön H-G, Kaufmann F, Kemper N, Andersson R. First approach to validate a scoring system to assess footpad dermatitis in Pekin ducks. *Europ Poult Sci.* 2019;83.

17. Knafo SE, Graham JE, Barton BA. Intravenous and intraosseous regional limb perfusion of ceftiofur sodium in an avian model. *Am J Vet Res.* 2019; 80:539–546.

18. Li C, Lesuisse J, Schallier S, Clímaco W, Wang Y, Bautil A, Everaert N, Buyse J. The effects of a reduced balanced protein diet on litter moisture, pododermatitis and feather condition of female broiler breeders over three generations. *Animal.* 2018;12:1493–1500.

19. Marques MVR, de Resende JS, Donatli RV, da Rocha Vilela DA, Ecco R, da Silva Martins NR. A bumblefoot outbreak and fatal septicemia in captive aquatic birds in Brazil. *Cienc Rural.* 2009;39:1905–1907.

20. Mayne RK. A review of the aetiology and possible causative factors of foot pad dermatitis in growing turkeys and broilers. *World Poultry Sci J.* 2005;61:256–267.

21. Nascimento CL, Ribeiro MS, Sellera FP, Dutra GHP, Simões A, Teixeira CR. Comparative study between photodynamic and antibiotic therapies for treatment of footpad dermatitis (bumblefoot) in Magellanic penguins (*Spheniscus magellanicus*). *Photodiagn Photodyn.* 2015;12:36–44.

22. Nielsen AMW, Nielsen SS, King CE, Bertelsen MF. Classification and prevalence of foot lesions in captive flamingos (Phoenicopteridae). *J Zoo Wildlife Med.* 2010;41:44–49.

23. Nielsen AMW, Nielsen SS, King CE, Bertelsen MF. Risk factors for development of foot lesions in captive flamingos (Phoenicopteridae). *J Zoo Wildlife Med.* 2012;43:744–749.

24. Olsen RH, Christensen H, Kabell S, Bisgaard M. Characterization of prevalent bacterial pathogens associated with pododermatitis in table egg layers. *Avian Pathol.* 2018;47:281–285.

25. Reidarson TH, McBain J, Burch L. A novel approach to the treatment of bumblefoot in penguins. *J Avian Med Surg.* 1999;13:124–127.

26. Reisfeld L, Barbirato M, Ippolito L, Cardoso RC, Nichi M, Sgai MGFG, Pizzutto CS. Reducing bumblefoot lesions in a group of captive Magellanic penguins (*Spheniscus magellanicus*) with the use of environmental enrichment. *Pesq Vet Bras.* 2013;33:791–795.

27. Reissig EC, Tompkins DM, Maloney RF, Sancha E, Wharton DA. Pododermatitis in captive-reared black stilts (*Himantopus novaezelandiae*). *J Zoo Wildlife Med.* 2011;42:408–413.

28. Remple J. A multifaceted approach to the treatment of bumblefoot in raptors. *J Exotic Pet Med.* 2006;15:49–55.

29. Sander S, Whittington JK, Bennett A, Burgdorf-Moisuk A, Mitchell MA. Advancement flap as a novel treatment for a pododermatitis lesion in a red-tailed hawk (*Buteo jamaicensis*). *J Avian Med Surg.* 2013;27:294–300.

30. Scagnelli AM. Manuka honey. *J Exotic Pet Med.* 2016;25:168–171.

31. Sellera FP, Sabino CP, Ribeiro MS, Fernandes LT, Pogliani FC, Teixeira CR, Dutra GHP, Nascimento CL. Photodynamic therapy for pododermatitis in penguins. *Zoo Biol.* 2014;33:353–356.

32. Shepherd EM, Fairchild BD, Ritz CW. Alternative bedding materials and litter depth impact litter moisture and footpad dermatitis. *J Appl Poult Res.* 2017;26:518–528.

33. Tolpinrud A, O'Brien MF, Justice WSM, Barrows M, Steele O, Gent S, Meredith AL. Infrared thermography as a diagnostic tool for pododermatitis in captive greater flamingos (*Phoenicopterus roseus*). *J Zoo Aquarium Res.* 2017;5:48–55.

34. Webb JK, Keller KA, Welle K, Allender MC. Evaluation of the inter- and intraindividual agreement of a pododermatitis scoring model in greater flamingos (*Phoenicopterus roseus*). *J Zoo Wildlife Med.* 2020;51:379–384.

35. Williams LK, Sait LC, Trantham EK, Cogan TA, Humphrey TJ. *Campylobacter* infection has different outcomes in fast- and slow-growing broiler chickens. *Avian Dis.* 2013; 57:238–241.

36. Wyss F, Wenker C, Hoby S, von Houwald F, Schumacher V, Doherr MG, Robert N. The effect of fine granular sand on pododermatitis in captive greater flamingos (*Phoenicopterus roseus*). *Anim Welfare.* 2014;23:57–61.

37. Wyss F, Wolf P, Wenker C, Hoby S, Schumacher V, Bechet A, Robert N, Liesegang A. Comparison of plasma vitamin A and E, copper and zinc levels in free-ranging and captive greater flamingos (*Phoenicopterus roseus*) and their relation to pododermatitis. *J Anim Physiol Anim Nutr.* 2014;98:1102–1109.

38. Wyss F, Schumacher V, Wenker C, Hoby S, Gobeli S, Arnaud A, Engels M, Friess M, Lange CE, Stoffel MH, Robert N. Pododermatitis in captive and free-ranging greater flamingos (*Phoenicopterus roseus*). *Vet Pathol.* 2015;52:1235–1242.

Speaker Information
(click the speaker's name to view other papers and abstracts submitted by this speaker)

Ellen Rasidi, BBiomedSc, BSc(Vet)(Hons), BVSc
Department of Conservation
Research and Veterinary Services
Wildlife Reserves
Singapore

URL: https://www.vin.com/doc/?id=9768895

The Avian Renal System from Form and Function to Clinical Perspectives

ExoticsCon Virtual 2020 Proceedings

Susan Orosz[1], DVM, PhD, DABVP (Avian), DECZM (Avian); M. Scott Echols[2,3,4], DVM, DABVP (Avian)

[1]Bird and Exotic Pet Wellness Center, Toledo, OH, USA; [2]Echols Veterinary Services, Salt Lake City, UT, USA; [3]Medical Center for Birds, Oakley, CA, USA; [4]Parrish Creek Veterinary Clinic, Centerville, UT, USA

Abstract

When discussing the urinary system and its unique aspects in birds, the important consideration is how do birds osmoregulate? All animals need to have adequate water in their extracellular fluid compartment to regulate cellular functions. In order to achieve water and ion homeostasis, animals range from osmoconformers to osmoregulators of their internal environment. Birds are osmoregulators.

Birds fall between mammals and other non-mammalian vertebrates for the organs of osmoregulation. Many avian species have salt glands that perform important osmoregulatory function, particularly in a marine environment. Like reptiles, birds utilize the lower portion of the GI tract as an important component for osmoregulation. This first part of the lecture for this discussion today will focus on the gross anatomy of the kidneys, including an overview of the blood supply in the area along with nerves. Based on this anatomy, it will include radiology and CT images of the kidneys.

The Kidney

External Anatomy

The renal system consists of the kidneys, ureters, and the urodeal portion of the cloaca. The kidneys of birds are fixed in ventral depressions of the synsacrum called the renal fossae, and they are symmetrical and retroperitoneal. They extend from the proximal end of the synsacrum just distal to the caudal extent of the lungs to the cranial end of the synsacrum. There can be diverticula of the abdominal air sacs that extend between the kidneys and the pelvis.

Unlike mammalian kidneys, the kidneys of birds are not divided into lobes grossly, but into divisions. Birds tend to have three divisions: cranial, middle, and caudal. The external iliac artery cranially and the ischiatic artery caudally run through the kidney so that it forms the three divisions. The portion of the kidney cranial to the external iliac artery is the cranial division; the portion between the two arteries is the middle; and that portion caudal to the ischiatic artery is the caudal division. In chickens and many of the parrot species, these divisions are distinct. The middle division in passerines is not distinct, as this portion blends with the cranial and more often the caudal division. In puffins, penguins, and herons, the caudal divisions are fused in the midline.

The spinal nerves from the lumbar and sacral plexuses move through the substance of the kidney to innervate their respective muscles to the lower body wall and the leg or are sensory to that area of the body. This is important clinically, as swelling of the kidneys causes reduction in nerve function, and the patient may present with inability to use the leg. This more often occurs in budgerigars, and they are presented to the veterinarian for a possible leg fracture.

In the domestic fowl, each division has on its surface slightly rounded projections that represent the renal lobules as they reach the surface. In general, the basic unit of the avian kidney is the lobule. They tend to be pear-shaped and wedged between the interlobular veins of the renal portal system. The wide, superficial portion represents the cortical region of the lobule and contains nephrons of both the cortical- and medullary-type nephrons.

The artery that supplies each lobule is found in its center with radiating branches. The drainage of the lobule consists of a vein that is intralobular like the artery. The collecting ducts of birds are found on the periphery of each lobule, which has been likened to the outer staves of a barrel. This is the opposite of mammals where the collecting ducts are intralobular and the arteries are interlobular.

The Arterial Supply to the Kidneys

The cranial, middle, and caudal divisions of the kidney are supplied by their respective arteries that arise from the abdominal aorta. The caudal division has a branch off the aorta that bifurcates on the surface of the kidney, resulting in two arteries that enter the parenchyma of this larger division.[1] These arteries then branch within the substance of the kidneys to eventually form the intralobular arteries. The intralobular arteries are found in the lobule approximately halfway between the inter- and intralobular veins.

The intralobular arteries form the short afferent glomerular arteries that immediately form the glomerulus. Each of these capillary tufts is simpler and much smaller than mammalian ones, as they consist of only 2–3 capillaries that have limited interconnections compared with mammalian capillary tufts. The capillary tufts of the loopless nephrons can consist of only one capillary loop that such an afferent arteriole forms a single capillary loop and then widens to form an efferent arteriole.

The arterial supply continues as efferent glomerular arterioles. These arterioles divide to form the second capillary plexus, the peritubular capillary plexus, which surrounds the epithelium of the convoluted tubules in the cortical region. In the medullary region, these arterioles from the efferent glomerular arterioles form arteriolar recta that lie alongside the loops of Henle before forming the venulae recta that drains the area.

The Venous Supply to the Kidneys

Blood from the area around the loops of Henle are drained by the venulae recta into the intralobular veins. The intralobular veins drain blood into the efferent renal veins or branches. Depending on the division of the kidney, the blood then flows into renal veins of its appropriate division. In the cranial division of the kidney, there can be several cranial renal vessels that then drain into the common iliac vein after the renal portal valve or into the abdominal vena cava directly. The caudal renal vein drains the middle and caudal divisions of the kidneys. Blood from the intralobular veins drains into the efferent renal veins or branches before draining into the caudal renal vein. The caudal renal vein, like the cranial renal vein, empties into the common iliac vein after the renal portal valve.

The renal portal system is involved in the secretion of urates in the blood so that it can be excreted by the kidneys in birds. The renal portal system supplies venous blood to the peritubular capillary plexuses that surround the proximal convoluted tubules at the periphery of the lobule. These tubules are responsible for the secretion of the urates into its lumen from these blood vessels. Urates are also filtered by the glomerulus, but its rate is insufficient. It has been suggested that the renal portal system provides about two-thirds of the blood supply to the kidneys that bypasses the glomeruli. This has important ramifications for the removal of urates along with other components, including drugs.

The renal portal system forms a venous ring with both kidneys. It consists of the right and left cranial and caudal renal portal veins. At the cranial end, the right and left cranial renal portal veins are connected via the internal vertebral venous sinus, which drains the vertebral column. The caudal end of the ring is completed by its anastomoses with the caudal mesenteric vein. There are numerous small venous branches that take origin from this ring to penetrate the parenchyma of the kidney. Each afferent renal branch has a muscular sphincter at its base to control the volume of blood that enters the substance of the kidney. These afferent renal venous branches are connected with the interlobular veins, which are connected to the peritubular capillary plexus at the periphery of each lobule.

In addition to the valves in the afferent renal venous branches, there are renal portal valves that control blood flow as well. These valves are found within the lumen of the common iliac veins between the renal and the portal veins. Blood can enter the portal venous ring from the external iliac veins, the internal iliac veins, the caudal mesenteric vein, and the ischiatic veins. Each valve is innervated by adrenergic and acetylcholine receptors. If the valves are open, then blood will flow directly into the caudal vena cava and not the substance of the kidney. Valve closure is inhibited by norepinephrine and epinephrine so that with "flight" there is available venous return to the heart. When the valve is closed, blood flows into the parenchyma of the kidney. Acetylcholine stimulates its closure.

From a clinical perspective, understanding these connections with the renal portal system helps explain the spread of neoplasia and infectious agents to other parts of the body. Blood from the renal portal system can flow through the portal valve into the vena cava, into the caudal mesenteric vein to the liver, and/or into the internal vertebral venous plexus within the vertebral canal. This shunt can result in the bypass of the kidney completely but often is only partially activated so that blood may go a variety of directions.

Diagnostics

Characterizing renal disease can be challenging and is often best performed by using a combination of diagnostic methods. Some diagnostics rely in blood and urine chemistry values such as uric acid and urinary gamma-glutamyl transferase (GGT), respectively. A thorough understanding of avian renal anatomy is needed when considering radiology, ultrasound, endoscopy, and advanced imaging (especially computed tomography [CT]).

A lateral view is the best method to radiographically view the kidneys. As viewed with a lateral radiograph, the absence of the normal dorsal diverticulum of the abdominal air sac (dorsal to the kidney and ventral to the synsacrum) may indicate renal enlargement. Also, **most** normal-sized kidneys do not extend ventral to an imaginary horizontal line (which is parallel to the spine) that passes through the ventral border of the acetabulum. Improper positioning can artifactually change the appearance of this air-filled diverticulum. Because the renal silhouettes are superimposed on a lateral view of the abdomen, an oblique view may also be used to distinguish each kidney. Renal density and gross size changes may indicate renal disease.

A barium series on the ventrodorsal view may also be helpful in identifying severe renal swelling. In normal birds, the angle between the proventriculus and the ventriculus relative to the spine should be about 45 degrees. With massive renal enlargement and lateral displacement of the proventriculus, and sometimes ventriculus, this angle increases.

Due to the presence of surrounding air sacs (ventrally) and bone (dorsally and laterally), ultrasonographic imaging of normal avian kidneys is difficult. In one study of 386 mixed bird species that underwent ultrasonographic evaluation of the urogenital tract, abnormalities such as renal cysts (6), cancer (12), and inflammatory nephromegaly (11) were identified in only 29 patients. The authors concluded that sonographic imaging of the normal kidney was not possible. Some disease conditions that either obliterate the air sacs or result in fluid accumulation in the coelomic cavity may actually improve renal ultrasonographic imaging. Renal tumors can be heterogenous compared to homogeneous with normal tissue, and they have no discernible border with the synsacrum dorsally as viewed with ultrasound. In these abnormal situations, ultrasonography can serve as a noninvasive and safe means to evaluate coelomic structures such as the kidneys. Also, contrary to statements above, some skilled ultrasonographers can position birds such that the kidneys are visible via ultrasound even if there is no pathology present.

Intravenous excretory urography has been described in birds as a method to gain information on kidney size, shape, and function. Lumeij reports using organic iodine compounds given IV at 2 mg/kg in the basilic vein. The organic iodine can be visualized radiographically in the heart and pulmonary artery within 10 seconds and outlining the kidneys and ureters 20 to 50 seconds later. After 2 to 5 minutes, the cloaca will be outlined. This technique should not be used in birds with severe renal compromise.

Intravenous excretory urography may have some limited uses in a clinical setting as demonstrated in the case report below. Dennis and Bennett successfully used a water-soluble, iodinated contrast agent (Renografin-76, Squibb Diagnostics, Princeton, NJ, USA) to evaluate the ureters post-ureterotomy in a double yellow-headed amazon parrot (*Amazona ochrocephala*). The agent was dosed at 400 mg/kg and was given in the right medial metatarsal vein. Radiographic images were taken at 1, 2, 7, and 10 minutes post-injection. Ureter peristaltic movement and size were successfully evaluated using this technique.

However, excretory urography combined with CT can provide significant information about the avian kidney. CT (computed tomography) is essentially a series of x-rays compiled to form 3D images. High-quality CT scans (generally 200-µm slice thickness or less) can provide unparalleled 3D information about kidneys in health and disease. MSE uses the following protocol for contrast CT studies in birds:

- First: Whole-body or region of interest (ROI) plain CT scan.
- Second: 4–6 ml/kg of Isovue 370 IV delivered over 2 minutes. Thirty seconds into the contrast perfusion, start the second whole-body or ROI CT scan.
- Third: Complete a last whole-body or ROI scan immediately after completing the contrast perfusion.

This form of CT angiography can demonstrate not only anatomic features of the kidney, but it also gives information about renal contrast clearance. Tumors, cysts, vasculature, and more can be highlighted by the contrast study giving anatomic information. In a normal (third) post-contrast CT scan, the ureters and cloaca fill with contrast. Lack of contrast filling in the ureters gives information about renal clearance and patency of the ureters.

When history, physical examination, and/or laboratory abnormalities support the presence of renal disease, consider biopsy. Currently, the only means to definitively diagnose avian renal disease and specific pathologic patterns is with a kidney biopsy and histopathologic evaluation. A renal biopsy is most frequently performed during endoscopic examination of the coelomic cavity and specifically, kidneys. Alternatively, the kidneys can be approached dorsally by cutting through the synsacrum. This dorsal approach obviously requires cutting through bone and increases the risk of damaging the more dorsally located lumbar and sacral nerve plexuses. Before a renal biopsy is performed, the cost–benefit analysis of the surgical procedure versus conservative therapy must be considered, as many birds have compromised health, especially if they have kidney disease.

In a study of 89 free-living birds of prey, 126 endoscopic coelomic-approach renal biopsy samples (two biopsies from 37 birds) using 1.8-mm biopsy forceps were taken. Post-biopsy hemorrhage averaged 67 seconds (10–172 seconds). The average biopsy was 2.2 mm long, 1.3 mm wide, and 1.0 mm deep. All samples contained proximal and distal tubuli and one to 89 glomeruli, with most having 25–29 glomeruli per histologic slide. One hundred thirteen of 126 samples could be evaluated well or very well. Sixty-six samples revealed lesions including subcapsular bleeding (19/66), inflammation (16/66), cell casts (12/66), periodic acid–Schiff-positive reactions (8/66), and protein casts (6/66). Correlation between endoscopically visible change and histologic disease was 76.1% (96/126). The cranial division was considered the best site to collect biopsy samples due to its size and visibility. The authors, Müller *et al.*, noted it was possible to obtain specimens from the middle and caudal renal divisions in larger birds.

Renal histologic lesions are rarely pathognomonic for a specific disease process. Many different diseases cause similar renal lesions. Additionally, different pathologists may make differing morphologic diagnoses on the same renal tissue. The author (MSE) encourages veterinarians to work with a pathologist familiar with normal and abnormal avian histology. Oftentimes, it is the pathologist's interpretation of a renal biopsy combined with the attending veterinarian's case familiarity that enables both parties to make a definitive diagnosis or build a reasonable differential diagnoses list compatible with the kidney lesions noted.

References

References are available upon request.

Speaker Information
(click the speaker's name to view other papers and abstracts submitted by this speaker)

M. Scott Echols, DVM, DABVP (Avian)
Echols Veterinary Services
Salt Lake City, UT, USA

Susan Orosz, DVM, PhD, DABVP (Avian), DECZM (Avian)
Bird and Exotic Pet Wellness Center
Toledo, OH, USA

URL: https://www.vin.com/doc/?id=9768900

Parrot Bornavirus: Current Clinical Summary of Diagnosis, Treatment, and Control

ExoticsCon Virtual 2020 Proceedings

Jeffrey M.B. Musser, DVM, PhD, DABVP (Dairy); J. Jill Heatley, DVM, MS, DABVP, DACZM

Department of Veterinary Pathobiology and Department of Small Animal Clinical Sciences, College of Veterinary Medicine and Biomedical Sciences, Texas A&M University, College Station, TX, USA

Introduction

Avian bornavirus (ABV) has a worldwide distribution, with reports from North America, Europe, South Africa, Japan, Sweden, and Brazil, to name a few. Avian bornavirus can cause proventricular dilatation disease; however, proventricular dilatation disease is really a misnomer, as many systems in the body may be affected by ABV—in particular the nervous system. Not all birds infected with ABV develop signs of PDD, and not all birds experiencing the syndrome PDD shed the virus or have other evidence of ABV infection. Due to the neurological involvement, clinical signs can include blindness, ataxia, weakness, weight loss, inability to perch, or passage of undigested food—with the bird eventually succumbing to opportunist gastrointestinal infections or starvation due to inability to ingest and digest food. While classical clinical signs of ABV infection include passage of undigested seed, fetid stool, vomiting undigested seed, weight loss, and a variety of neurologic signs, many other diseases can cause these same clinical signs. Differentials that should be considered for PDD include heavy metal intoxication, occult parasitism, fungal or bacterial infection, and a variety of other causes of encephalitis.

Avian Bornavirus Infections

Avian bornavirus infection is caused by an enveloped single-strand RNA virus in the family *Bornaviridae*. There are five species of *Bornaviridae* that infect birds: *Psittaciform 1 bornavirus*; *Psittaciform 2 bornavirus*; *Passeriform 1 bornavirus*; *Passeriform 2 bornavirus*; and *Waterbird 1 bornavirus*. Within each species, there are numerous genotypes that can infect a variety of different bird species (Table 1).

Table 1. *Bornaviridae* species, genotypes, and animals infected (adapted from Escandon P. Thesis. 2020)

Species	Genotypes	Animals infected
Psittaciform 1 bornavirus	Parrot bornavirus (PaBV) 1,2,3,4,7,8	Psittacines including: African grey parrots (*Psittacus* spp.) Amazon parrots (*Amazonas* spp.) Salmon-crested cockatoo (*Cacatua moluccensis*) Macaws (*Ara macao*, *Ara* sp.) Cockatiels (*Nymphicus hollandicus*)
Psittaciform 2 bornavirus	Parrot bornavirus (PaBV-5)	Psittacines including: Palm cockatoo (*Probosciger aterrimus*)
Passeriform 1 bornavirus	Canary bornavirus (CnBV) 1,2,3	Canary (*Serinus canaria*)
Passeriform 2 bornavirus	Munia bornavirus 1 (MuBV-1) Estrildid finch bornavirus 1 (EsBV-1)	Munia finches (*Lonchura punctulata*) Estrildid finches (Estrildidae)
Waterbird 1 bornavirus	Aquatic bird bornavirus 1 (ABBV-1)	Canada geese (*Branta canadensis*) Herring gulls (*Larus argentatus*) Ring-billed gulls (*Larus delawarensis*) Laughing gulls (*Leucophaeus atricilla*) Emu (*Dromaius novaehollandiae*)
Elapid 1 bornavirus	Loveridge garter snake bornavirus (LGSV-1)	Loveridge garter snakes (*Elapsoidea loveridgei*)
Mammalian 1 bornavirus	Borna disease virus (BoDV) 1,2	Horses, sheep, bicolored shrew, cats

While transmission and pathogenesis details of the disease in companion birds are still under active investigation, infection rate may be as high as 15%. The incubation period and shedding during this period is unknown; however, shedding by asymptomatic birds is likely. Infected birds are thought to shed the virus for life; however, there is a report of a captive flock of ABV-infected cockatiels spontaneously eliminating the virus, becoming PCR negative for ABV. Viral RNA and infectious virus have been found in urine, feces, tears, and oral secretions. This virus is likely transmitted by aerosol, contact or ingestion, and vertical methods; however, at present, the only experimental means of infecting a bird is through intramuscular, intranasal, and ocular administration of the virus.

Diagnosis

Diagnosis is still an active area of investigation. Avian bornavirus is mainly a neurotropic virus but can be found in many tissues and cell types. Diagnostic techniques include molecular techniques by reverse-transcription polymerase chain reaction (RT-PCR), serological assays such as Western blot, enzyme-linked immunosorbent assays (ELISA), indirect enzyme-linked immunosorbent assay and immunohistochemistry, and virus isolation. ABV antigens and genetic material have been found in the brain, eye, spinal cord, heart, gastrointestinal system, adrenal glands, kidneys, urine, feces, tears, oral secretions, and feather dander. To the authors' knowledge, there are presently no commercially available rapid immunological assays available. Radiographs, CT or fluoroscopy with contrast may also be useful in these cases. When present, gross lesions consist of dilated proventriculus and cardiac enlargement. Histopathology of proventricular dilatation disease is characterized by nonsuppurative inflammation in the central, peripheral, and autonomic nervous systems. Immunohistochemistry of these same lesions often reveals viral antigen.

Treatment

To our knowledge, no specific antiviral treatment has been effective at either eliminating virus from the host, reducing viral shedding, or ameliorating clinical sigs of avian bornaviral infection. Ribavirin, a combination of ribavirin and interferon-alpha, and favipiravir have shown to interfere with avian bornavirus replication in cell culture but have not been evaluated in an infected bird.

The use of nonsteroidal anti-inflammatory drugs has been suggested as a treatment for birds with clinical signs of ABV infection. However, with experimentally infected cockatiels, neither meloxicam nor celecoxib altered clinical presentation, viral shedding, gross lesions, histopathology, or viral distribution. In fact, treatment with NSAIDs may have induced gastrointestinal toxicity resulting in potential intestinal bleeding. Thus, caution is suggested in using long-term NSAID treatment. Treatments with steroids and other immunosuppressive drugs have been suggested, but the authors could find no published studies. Thus, supportive treatment of clinically infected birds is presently the only viable option.

Prevention

Prevention involves screening for viral infection and shedding via PCR. Maintenance of an ABV-free collection is preferred but is rarely currently feasible. Testing and separation are difficult based on intermittent shedding and latency of virus.

No recommended commercially available vaccines are available. Work is progressing on vectored, subunit, and killed vaccines. Vaccination has not been shown to protect against infection nor arrest virus shedding. It may result in decreased severity of clinical signs, but the literature is not unanimous on this point. A major hindrance in vaccine research is the development of an experimental model that reflects infection in the wild. All the vaccine-challenge studies have utilized intramuscular, intranasal, or ocular administration of the virus.

Although appropriate disinfection remains uninvestigated, ABV is an enveloped virus, and enveloped viruses are likely easily inactivated. Thus, most of the disinfectants used in aviaries would be suitable, though this is only speculation. As to ABV's stability in the environment, preliminary studies suggest that the virus is stable between pH 7 to 10, but it is rapidly inactivated outside this range. At 26°C and 21% humidity, ABV remains infectious for only 24 hours; but after 48 hours, 99.97% of the virus is inactivated, and after four days, no infectious virus is present. Further studies need to be done on disinfection and sentimental stability of ABV.

Research

The Schubot Exotic Bird Health Laboratories have tested >500 free-living, non-Psittaciformes of more than 30 species (Table 2) for avian bornavirus (ABV). Birds sampled include hunter-killed ducks; raptors and aquatic birds that died or were euthanized during rehabilitation; birds that were culled by animal control authorities; Passeriformes that were collected for museum study; and rehabilitated pre-release ducks and pelicans from Wildlife Center of Texas. Samples tested include choanal and cloacal swabs and brain, liver, and splenic tissues. Avian bornavirus sequences were amplified using two multiplexed primer sets. One recognized matrix protein (M protein) genes, and the other recognized conserved regions of the nucleoprotein (N protein) genes. Positive results were verified by repeat PCR testing, sequencing, and culture. Many non-psittaciform species were identified with avian bornavirus, particularly Anseriformes and Charadriiformes; however, sample sizes in other avian taxa were low. An improved recovery rate for ABV from the brain occurred compared with a very low recovery rate from swabs from choana, cloaca, or droppings. Clinical signs attributable to ABV in psittacine birds (gastrointestinal tract dysfunction, neurologic signs) are seldom noted. Sequencing has revealed multiple ABV isolates that are not closely related to the isolate that appears most likely to cause clinical signs of Psittaciformes (ABV4).

Table 2. Non-Psittaciformes tested by Schubot Exotic Bird Health Laboratories for avian bornavirus

Species group	Brain	Choanal, cloacal, dropping swab	Kidney, liver, spleen	Positive samples	Area
Anseriformes	625	60	0	76	TX, NJ, KS
Passeriformes	9	-	73	0	TX/Africa
Charadriiformes	47	-	0	6	NY, NJ, NH
Raptors	31	-	-	0	TX
Coraciiformes	-	-	3	0	Africa
Piciformes	-	-	7	0	Africa
Pelecaniformes	10	12	-	0	TX
Columbiformes	8	-	-	0	TX

From 2009 to 2013, after the discovery of ABV and fulfillment of Koch's postulates as the causative agent of proventricular dilatation disease, Schubot Exotic Bird Health Center started accepting samples from private practice and zoo veterinarians to assess parrots for ABV. Most samples were choanal/cloacal, choanal or fecal swabs obtained from live parrots. Over 460 samples have been submitted and screened via PCR for the M protein and N protein genes. Macaw species were the most commonly submitted (n=133) followed by cockatoos (n=92), African grey parrots (n=52), and *Amazona* spp. (n=39). Prevalence based on positive PCR results was approximately 15% across all species. Most taxa were similar in prevalence, although *Amazona* spp. (10%, 4/35) and Macaws (*Ara* spp.) (13%, 17/133) were lower, and Australian species were higher (cockatoos 18% [17/92] and other Australian species 21% [7/26]). Cockatiels (9/35), Eclectus (7/21), African grey parrots (8/52), blue-throated macaws (2/8), and sun conures (3/14) appear to be the species with the highest likelihood of testing PCR positive for ABV, although our sample sizes are small. Many parrot species have yet to have a positive PCR for ABV in our lab, but our sample sizes are below 10 for all of these species. Future recommended epidemiologic research in ABV for parrots should include identification of the preferred sample type to detect ABV (swab, urine, or feces) in the clinically ill bird versus the healthy bird, and gender and seasonal influences on viral shedding.

Take-Home Points

1. ABV infection is widespread in captive and free-living birds; many infected birds are asymptomatic and do not have neurologic or other clinical signs.
2. The diagnosis of PDD can be challenging, but all efforts to rule in or out ABV infection as well as other causes of GI tract or neurologic illness should be included in a case workup.
3. The diagnosis of ABV infection should not be used as a sole reason for euthanasia.
4. Birds suffering from PDD may or may not be infected or shedding ABV; latency and immune-mediated aspects may play a role in disease occurrence.

5. Means of transmission have yet to be fully elucidated and should be further researched.

6. Efficacious disinfection protocols and environmental stability of the virus have not been fully researched.

References

1. de Araujo JL, Rodrigues-Hoffmann A, Giaretta PR, Guo J, Heatley J, Tizard I, Rech RR. Distribution of viral antigen and inflammatory lesions in the central nervous system of cockatiels (*Nymphicus hollandicus*) experimentally infected with parrot bornavirus 2. *Veterinary Pathology*. 2019;56(1):106–117.

2. de Araujo JL, Hameed SS, Tizard I, Escandon P, Giaretta PR, Heatley JJ, Hoppes S, Rech RR. Cardiac lesions of natural and experimental infection by parrot bornaviruses. *Journal of Comparative Pathology*. 2020;174:104–112.

3. Boatright-Horowitz SL. Avian bornaviral ganglioneuritis: current debates and unanswered questions. *Veterinary Medicine International*. 2020.

4. Delnatte P, Berkvens C, Kummrow M, Smith DA, Campbell D, Crawshaw G, Ojkic D, DeLay J. New genotype of avian bornavirus in wild geese and trumpeter swans in Canada. *Veterinary Record*. 2011;169:108.

5. Escandon P, Heatley JJ, Tizard I, Guo J, Shivaprasad HL, Musser JM. Treatment with nonsteroidal anti-inflammatory drugs fails to ameliorate pathology in cockatiels experimentally infected with parrot bornavirus-2. *Veterinary Medicine: Research and Reports*. 2019;10:185.

6. Escandon P, Heatley JJ, Berghman LR, Tizard I, Musser JM. Comparison of four anti-avian IgY secondary antibodies used in Western blot and dot-blot ELISA to detect avian bornavirus antibodies in four different bird species. *Veterinary Medicine: Research and Reports*. 2019;10:141–150.

7. Guo J, Payne S, Zhang S, Turner D, Tizard I, Suchodolski P. Avian bornaviruses: diagnosis, isolation, and genotyping. *Current Protocols in Microbiology*. 2014;34:15I.1.1–33.

8. Fluck A, Enderlein D, Piepenbring A, Heffels-Redmann U, Herzog S, Pieper K, Herden C, Lierz M. Correlation of avian bornavirus-specific antibodies and viral ribonucleic acid shedding with neurological signs and feather-damaging behaviour in psittacine birds. *Veterinary Record*. 2019;184(15):476. doi: 10.1136/vr.104860.

9. Gray P, Hoppes S, Suchodolski P, *et al*. Use of avian bornavirus isolates to induce proventricular dilatation disease in conures. *Emerging Infectious Diseases*. 2010;16(6):473–479.

10. Honkavuori KS, Shivaprasad HL, Williams BL, Quan PL, Hornig M, *et al*. Novel bornavirus in psittacine birds with proventricular dilatation disease. *Emerging Infectious Diseases*. 2008;14(12):1883–1886.

11. Hoppes S, Gray PL, Payne S, Shivaprasad HL, Tizard I. The isolation, pathogenesis, diagnosis, transmission, and control of avian bornavirus and proventricular dilatation disease. *Veterinary Clinics of North America: Exotic Animal Practice*. 2010;13(3):495–508.

12. Kistler AL, Gancz A, Clubb S, Skewes-Cox P, *et al*. Recovery of divergent avian bornaviruses from cases of proventricular dilatation disease: identification of a candidate etiologic agent. *Virology Journal*. 2008;5:88.

13. Kuhn JH, Dürrwald R, Bào Y, Briese T, Carbone K, Clawson AN, Derisi JL, Garten W, Jahrling PB, Kolodziejek J, Rubbenstroth D. Taxonomic reorganization of the family Bornaviridae. *Archives of Virology*. 2015;160(2):621–632.

14. Maclachlan NJ, Dubovi EJ, eds. Bornaviruses, subsection avian bornaviruses. In: *Fenner's Veterinary Virology*. Academic Press; 2010.

15. Murray O, Turner D, Streeter K, Guo J, Shivaprasad HL, Payne S, Tizard I. Apparent resolution of parrot bornavirus infection in cockatiels (*Nymphicus hollandicus*). *Veterinary Medicine: Research and Reports*. 2017;8:31–36.

16. Niller HH, Angstwurm K, Rubbenstroth D, Schlottau K, Ebinger A, Giese S, Wunderlich S, Banas B, Forth LF, Hoffmann D, Höper D. Zoonotic spillover infections with Borna disease virus 1 leading to fatal human encephalitis, 1999–2019: an epidemiological investigation. *The Lancet Infectious Diseases*. 2020;20(4):467–477.

17. Payne S, Covaleda C, Jianhua G, Swafford S, Baroch J, Ferro PJ, *et al*. Detection and characterization of a distinct Bornavirus lineage from healthy Canada geese (*Branta canadensis*) *Journal of Virology*. 2011;85:12053–12056.

18. Payne S, Shivaprasad HL, Mirhosseini N, Gray P, Hoppes S, Weissenbock H, Tizard I. Unusual and severe lesions of proventricular dilatation disease in cockatiels (*Nymphicus hollandicus*) acting as healthy carriers of avian bornavirus (ABV) and subsequently infected with a virulent strain of ABV. *Avian Pathology*. 2011;40(1):15–22.

19. Pinto MC, Craveiro H, Johansson Wensman J, Carvalheira J, Berg M, Thompson G. Bornaviruses in naturally infected *Psittacus erithacus* in Portugal: insights of molecular epidemiology and ecology. *Infection Ecology & Epidemiology*. 2019;9(1):1685632.

20. Pinto MC, Rondahl V, Berg M, Ågren E, Carvalheira J, Thompson G, Johansson Wensman J. Detection and phylogenetic analysis of parrot bornavirus 4 identified from a Swedish blue-winged macaw (*Primolius maracana*) with unusual nonsuppurative myositis. *Infection Ecology & Epidemiology*. 2019;9(1):1547097.

21. Rall I, Amann R, Malberg S, Herden C, Rubbenstroth D. Recombinant modified vaccinia virus Ankara (MVA) vaccines efficiently protect cockatiels against parrot bornavirus infection and proventricular dilatation disease. *Viruses*. 2019;11(12):1130.

22. Weissenbock H, Sekulin K, Bakonyi T, Hogler S, Nowotny N. Novel avian bornavirus in a nonpsittacine species (canary; *Serinus canaria*) with enteric ganglioneuritis and encephalitis. *Journal of Virology*. 2009;83(21):11367–11371.

Speaker Information
(click the speaker's name to view other papers and abstracts submitted by this speaker)

Jeffrey M.B. Musser, DVM, PhD, DABVP (Dairy)
Department of Veterinary Pathobiology and
Department of Small Animal Clinical Sciences
College of Veterinary Medicine & Biomedical Sciences
Texas A&M University
College Station, TX, USA

J. Jill Heatley, DVM, MS, DABVP, DACZM
Department of Small Animal Clinical Sciences
College of Veterinary Medicine and Biomedical Sciences
Texas A&M University
College Station, TX, USA

URL: https://www.vin.com/doc/?id=9768897

The ADR Backyard Hen

ExoticsCon Virtual 2020 Proceedings

Renée Schott, DVM, CWR
The Wildlife Rehabilitation Center of Minnesota, Roseville, MN, USA

Abstract

Backyard hens are continuing to grow in popularity for their appeal as a local source of fresh eggs. However, they often become a beloved pet and part of the family. When these hens present for being sick, families want veterinary care but are often not prepared to spend as much as for other companion species. This masterclass will review the most common presentations for sick pet hens and present some quick diagnostics to help guide owners' decisions in treatment options.

Introduction

Backyard chickens are increasingly being allowed in cities, and clients are keeping them for fun and eggs. Urban clients will often develop strong emotional bonds and consider them pets on the same level as their companion cats and dogs. However, preventative care is not common, so these pets will usually present very sick without any prior medical history. Chickens are adept at hiding signs of disease, and novice keepers will likely miss subtle clinical signs of disease[1]; because of this, sick chickens should be considered an emergency. Please note: Many "traditional" infectious diseases are **not** going to be covered in this class, as there is a plethora of literature, treatments, and quarantine recommendations from the production poultry literature. This talk will be focusing on the hen with generalized symptoms, which often presents on emergency.

Veterinary Examination and Diagnostics

History

Collect all history from the owner—may be a clue to the current condition or can spur a husbandry conversation for after the crisis. Be sure to ask the following:

- **Signalment**: Age, breed, standard or bantam
- **Acquisition**:
    - When was the bird obtained?
    - From private party or hatchery? Did the previous owner get from a hatchery?
    - If from a hatchery, were any vaccinations given?
    - How long (and where) did the owner quarantine after obtaining?
- **Husbandry** (ideally house visit, or ask for pictures):
    - Winter housing (temperature, humidity, substrate, roosts, nest boxes, frequency of cleaning, airflow, etc.)
    - Summer housing (temperature, humidity, substrate, roosts, nest boxes, frequency of cleaning, airflow, etc.)
    - Free range, fenced, netting top, other?
    - How many birds in flock, age, sex, size, breed, and personalities; where is this bird in the pecking order?
    - Last time a new bird was added to flock
    - Food and water
        - Clean water available and changed daily? How many places? Out of sun/cold?
        - Fresh feed offered daily? How many places?
            - What kind of feed?
        - Calcium supplement offered?
        - Any other foods offered?
        - How often/how much treats? Table scraps, etc.?
- **Laying history**:
    - For how many years?
    - Frequency of egg laying?
    - Last time hen laid?
    - Size of egg?
    - Any abnormalities on the eggs (blood on shell, calcium deposits, abnormal shape or size, shell-less eggs)?
    - Recent broodiness?
- Prior medical issues?
- When did owner first start to notice something off with hen? What was noticed?
- Appetite?
- Feces/urates?
- Activity level?

Physical Exam

Perform as in any other species and pay attention for any of the following abnormalities:

- **Hands-off exam**: Watch for any dyspnea (tail bobbing, any obvious/exaggerated inhalation/exhalation, OMB, etc.)
- **Beak, mouth**: Look for any plaques, stringy saliva, color of mucous membranes, beak discoloration
- **Eyes**: Clear, open and bright? Or half-closed, sunken, dull? Ensure no swellings around eyes
- **Comb/wattle**: Color and size—ask owner if it is redder/larger or paler/smaller vs. when the hen is laying normally
- **Feathers/skin**: Look for lice, mites, any areas of feather loss or wounds? All feathers preened and out of sheaths? In good condition and fresh or worn and damaged? Stress bars?
- **Crop**: Is it full? Distended? Palpate, and what do you feel—feed? Hay? Mats? Liquid? Gas?
- **Auscultation**: Listen to heart for murmurs. Listen to each air sac individually, as each can insulate sounds from the others
- **Vent**: Clean? Feathers around vent full of feces? Diarrhea?
- **Caudal/ventral coelom**: Palpate—is it enlarged/distended? With fluid or soft tissue? Painful?
    - Ascites and/or coelomic distention is an incredibly common abnormal physical exam finding in the ADR hen

Two-View Radiographs

Ventrodorsal and lateral views are required with any ADR hen. They can quickly be done in a low-stress manner without anesthesia. Sedation (0.5 mg/kg midazolam IM ± 2 mg/kg butorphanol IM) is great to use in any hen, but especially necessary in stressed individuals or those in respiratory distress.

Evaluate the entire radiograph and be sure to evaluate the opacities of the caudal coelom and relative location of the ventriculus to the rest of the coelomic contents.

Ultrasound

Great for internal examination **and** for therapeutic coelomic centesis

Blood Lead Levels

Lead is in surprising places in the environment, and many chickens may have detectable blood lead levels. Any blood lead level above 10 µg/dl should be chelated, and the eggs should be discarded, as lead can pass into the egg and onto humans if eaten.[2]

Common Diagnoses

Secondary Comorbidities

Owners may come in with the following as a primary problem, but they are almost always a secondary comorbidity. Search and find the primary pathology!

- Gastrointestinal parasites
- Lice, mites
- Crop stasis
- Diarrhea of unknown origin

Primary Pathology

- Neoplasia: Is common in any area, but often will arise in the reproductive tract
    - Hard to diagnose; it is a diagnosis of exclusion usually
    - Exploratory celiotomy can sometimes reveal carcinomatosis
- Reproductive problems:
    - Ectopic eggs (yolks only or partially/completely shell) are when developing egg(s) enter the coelomic cavity through an oviductal rupture or reverse peristalsis from oviductal (masses, impactions, salpingitis), cloacal, or unknown reasons.[3] Ectopic eggs are often termed "internal laying" and may result in a sterile or non-sterile coelomitis.
    - Impacted oviduct: Salpingitis (sterile,[3] caused by *E. coli* or *E. coli* found secondarily[4]), cystic hyperplasia, metritis, neoplasia, or unknown etiologies may cause caseated material to form—often in layers around misshapen/ruptured/soft-shelled/partially formed egg pieces.[3] This material can cause oviduct impactions if it gets big enough. This material can be laid whole or in pieces if small enough and are often termed "lash eggs." Lash eggs are most common in older, prolific-laying breeds.
- (Less common) gout/urolithiasis[5]
    - Often due to laying feed fed to hens (or rooster) not in lay
    - Predisposing factors: Infectious bronchitis, prior renal damage, water deprivation
- (Less common) elevated blood lead levels: Chickens are stoic and often do not have overt clinical signs related to mildly elevated lead levels, but they likely have subclinical/underlying problems caused by the lead. Lead will also make recovery from any other condition more difficult, so it should be chelated whenever >10 µg/dl.

Uncommon Diagnoses

- Egg-binding, dystocia: Uncommon in poultry (occur in small bantams)[3]
- Primary crop impaction: Crop stasis or GIT stasis is almost always due to an underlying disease process, unless the hen is a meat breed

Treatments

Legalities

In the United States (and in many other countries[6]), all chickens are considered meat animals even if owned by a vegan on a farm sanctuary. It does not matter if the bird or eggs will never be used as food; in the U.S., simply being this species makes it a food animal in the eyes of the FDA.[7]

Below is the list of drugs that are prohibited from extralabel use; these drugs may never be used in chickens:

- Antivirals
- Cephalosporins
- Chloramphenicol
- Clenbuterol
- Diethylstilbestrol
- All fluoroquinolones, including enrofloxacin (any -oxacin)
- All glycopeptides, including vancomycin
- All nitroimidazoles, including metronidazole
- All nitrofurans, including nitrofurazone
- Indexed drugs
    - Example: Suprelorin F (Virbac)

FARAD has addressed egg residues.[7]

Supportive Care

Necessary with any diagnosis and will make an incredibly huge difference in the attitude of the hen.

- Subcutaneous fluids: 50 mg/kg/day (great to add in vitamin B complex 1–3 ml/L, and if patient has been inappetent for a few days, 30 mEq/L KCl). Monitor hydration BID and supplement SC fluids as needed.
- Analgesia: Recommended in all conditions. The author prefers meloxicam (1 mg/kg BID), recognizing this can make a gout case worse.
- Nutrition: Oftentimes these hens will be anorexic for some time. Switching to a higher-quality feed or moistening the food may entice hens to eat. Feel free to use insects (mealworms, earthworms, etc.) to entice, as well as fresh fruits and vegetables. Monitor weight daily, and if patient is losing weight or is anorexic, gavage feeding is necessary. If prolonged anorexia, start with recovery formulas for birds. If only off feed for a few days, moisten high-quality feed, blend, and gavage 30 ml/kg/feeding. Feeding BID is usually adequate.

Treatments of Specific Conditions

- Neoplasia:
    - Supportive care
    - Coelomocentesis if ascites; ultrasound can help guide
    - If legal in one's country, deslorelin may help slow the progression of some neoplasias,[8,9] but it is not legal to use in the U.S. for species other than ferrets for adrenal disease.
- Reproductive problems:
    - Supportive care

- If legal in one's country, deslorelin is implanted as soon as possible. This will put the hen out of lay,[10] stopping the problem from compounding and buying some time.
- Two weeks after implantation, exploratory celiotomy may be required to "clean out" any eggs laid internally, lash eggs, or other gross pathology.
- Owners must be warned that implantation with deslorelin is often needed for the rest of the hen's life, as most disorders will recur after the implant wears off.
- Salpingectomy may be needed for recurrent lash egg/salpingitis.

Conclusion

Unfortunately, the most common disorders in the aging backyard hen are chronic, terminal conditions that will require lifelong treatment, care, and possibly surgery. With a dedicated owner, these hens can go on to live years, but may not if neoplasia is resistant to deslorelin. Regardless, it is important to have a conversation with the owner to determine the relationship the hen has in the family; being open and honest about prognosis, cost, and treatment options from the first visit is important.

References

1. Crespo R, Senties-Cue G. Postmortem survey of disease conditions in backyard poultry. *J Exot Pet Med.* 2015;24:156–163.
2. Leibler JH, *et al.* Lead exposure to children from consumption of backyard chicken eggs. *Environ Res.* 2018;167:445–452.
3. Echols MS. Soft tissue surgery. In: Greenacre CB, Morishita TY, eds. *Backyard Poultry Medicine and Surgery.* Wiley Blackwell; 2015:220–262.
4. Morishita T, Porter R. Gastrointestinal and hepatic diseases. In: Greenacre CB, Morishita TY, eds. *Backyard Poultry Medicine and Surgery.* Wiley Blackwell; 2015:181–203.
5. Applegate TJ. Backyard poultry nutrition. In: Greenacre CB, Morishita TY, eds. *Backyard Poultry Medicine and Surgery.* Wiley Blackwell; 2015:72–81.
6. Whitehead ML, Roberts V. Backyard poultry: legislation, zoonoses and disease prevention. *J Small Anim Pract.* 2014;55:487–496.
7. Marmulak T, *et al.* FARAD Digest. Egg residue considerations during the treatment of backyard poultry. *J Am Vet Med Assoc.* 2015;247:1388–1395.
8. Keller KA, *et al.* Long-term management of ovarian neoplasia in two cockatiels (*Nymphicus hollandicus*). *J Avian Med Surg.* 2013;27:44–52.
9. Mans C, Pilny A. Use of GnRH-agonists for medical management of reproductive disorders in birds. *Vet Clin North Am Exot Anim Pract.* 2014;17:23–33.
10. Eusemann BK, *et al.* Influence of a sustained release deslorelin acetate implant on reproductive physiology and associated traits in laying hens. *Front Physiol.* 2018;9:1–11.

Speaker Information

(click the speaker's name to view other papers and abstracts submitted by this speaker)

Renée Schott, DVM, CWR
The Wildlife Rehabilitation Center of Minnesota
Roseville, MN, USA

URL: https://www.vin.com/doc/?id=9768899

ARAV

The Reptilian and Amphibian Circulatory Systems: Form, Function, and Venipuncture Sites

ExoticsCon Virtual 2020 Proceedings

Jeanette Wyneken, PhD
Department of Biological Sciences, Florida Atlantic University, Boca Raton, FL, USA

Abstract

The circulatory system is the functional delivery and distribution system for the body's heat, fluids, gases, nutrients, many ions, hormones, cellular waste products, and immune responses. Blood samples and analyses provide major insight into an animal's near-term health. This overview of the circulatory system provides the framework to understand the major anatomical patterns, the pathways blood takes, as well as the functional relationships among the structures in amphibian and reptilian patients.

Introduction

There are fundamental differences in amphibian and reptilian circulatory systems. Amphibians and reptiles have hearts formed of a thin-walled sinus venosus, and two atria. Amphibian hearts have a single ventricle with outflow to the paired aortae and a pulmocutaneous system that supports gas exchange. Reptiles have a muscular ventricle that is partially divided as cava in chelonians, lizards, snakes, and tuataras and fully divided into right and left ventricular chambers in crocodilians. The basic pattern of arterial circulation leaving the heart is paired aortae and the aortic arches. The left and right aortae join as a single vessel in the coelomic cavity and supply the kidneys, gonads, hindlimbs, and tail (when a tail is present).

Amphibian Circulation

The location of the heart varies within lineages and even within taxa. The amphibian heart tends to be located just cranial to the shoulders and extends to the base of the neck in Anurans (frogs and toads) and Caudata (salamanders and newts). Because frogs, mudpuppies, and urodele salamanders (especially axolotls) were used for decades in biology labs and developmental and reproductive studies, there is a fairly rich array of descriptive and illustrated guides available; many are in the older literature, yet remain as sound references.[1-5] There is relatively little information on the Gymnophiona (Caecilians).

Most amphibians undergo metamorphosis; consequently, the circulation of larval forms includes major branches of the aortic arches supplying the gills and tail. In anurans, both gills and tail regress, and limbs develop. Concurrent with these gross morphological changes is significant remodeling of the circulatory systems, particularly the aortic arches and the tributaries of the femoral arteries and caudal venous drainage. The changes in the larval to adult circulation in the Caudata usually are less extreme, largely due to the larval forms having legs and they retain a tail during metamorphosis. In most anurans and urodele larval forms, the paired aortae give off branches to three pairs of aortic arches to external gills and a pair of branches to each developing lung (the pulmonary arteries); cranial to the gills, the internal and external carotids extend into the head. The external carotid connects to the dorsal aorta. In adults, the pulmonary arteries and three aortic arches persist as systemic vessels, bypassing the now resorbed gills, and the external carotid loses its connection to the dorsal aorta. In adult frogs and toads, a cutaneous artery arises from the same aortic arch that gives rise to the pulmonary artery.[5] Venous drainage in salamanders is via a caudal vein that is joined by external iliac and femoral veins. Blood can then return to the heart via several alternative routes. It can flow from femoral veins to pelvic veins and join the single ventral abdominal vein, which travels cranially. The ventral abdominal vein is joined by the hepatic portal vein, and both enter the liver. Alternatively, flow from the hindlimbs can enter the external iliac veins, which flow into the renal portal veins, and then blood either enters the kidneys and leaves through the postcava to the sinus venous, or it continues cranially to the posterior cardinal veins where flow from the subclavian arteries and the external and internal jugular veins flow into the sinus venosus.[5]

Reptilian Circulation

The reptilian heart tends to be positioned roughly along the axial midline. In chelonians, the shoulder joints and in radiographs the girdles provide good landmarks; the heart is typically just caudal to the level of the acromion processes and cranial to the distal procoracoid process-procoracoid cartilage junctions. Part of the heart is often dorsal to margins of the humeral-pectoral scutes; however, in some species, it is positioned more posteriorly, along the midline between humeral-pectoral and pectoral-abdominal scutes lines.[10] In soft-shelled turtles (Trionychidae), the heart is displaced to the right.[6]

In lizards, the heart position can be almost in the gular regions (geckos of the genus *Hemidactylus* and bearded dragons, *Pogona vitticeps*) or more caudally within the rib cage (chamaeleonine, iguanine, and varanid lizards). In adult or large, immature iguanid and varanid lizards, the heart is located more caudally in the thorax; the caudal aspect of the ventricle is aligned with the level of elbow when the forelimbs are held retracted against the body. The heart may virtually fill much of the thoracic cavity in young lizards, especially *Pogona vitticeps*.

In snakes, the heart generally is located at a position that is 22–33% of the snout-vent length (15–25% of total length), caudal to the rostrum. However, the heart is often shifted caudally in totally aquatic snakes, to a position that is 25–45% of total length posterior to the rostrum.[6-8]

Reptilian cardiovascular anatomy varies with taxon. Typically, reptilian hearts are described by major structural patterns of organization and within taxonomic contexts. Because chambers of non-crocodilian ventricle normally are not morphologically partitioned into systemic and pulmonary circuits, structures are frequently described functionally by the relative condition of blood's oxygenation (high- or low-oxygen blood).

The heart is located within the pericardium. The pericardial sac typically contains clear pericardial fluid that, at least in turtles, is rich in calcium and magnesium ions. The alkaline pericardial fluid that bathes the heart may play a minor role in protection from acidosis during apnea.[9] The caudal aspect of the pericardium and apex of the ventricle are attached to the visceral peritoneum by a cord-like ligament, the gubernaculum cordis. This structure is present in most chelonians, most lizards, tuataras, and crocodilians.[6,10] When present, the structure anchors the ventricle so that longitudinal tension develops during contraction; this tension is one possible mechanism by which the muscular walls and ridges may functionally divide the otherwise incompletely divided ventricle during contraction, rather than retracting the ventricle craniad.[6] The gubernaculum cordis is not present in snakes or varanid lizards.[6]

The reptilian heart is more complex than its exterior suggests. The heart is composed of five or six functional chambers and compartments (blood flows into sinus venosus, right atria, and the compartments of the ventricle; then it either leaves through the pulmonary trunk to the lungs and via the paired aortae or during apnea, pulmonary flow is restricted and blood is routed to the system via the two aortae. Flow returning from lungs via the pulmonary vein enters the left atrium, which then leads to the ventricular compartment. There are three basic patterns of reptilian ventricular structure; most squamates (lizards and snakes), chelonians, and rhynchocephalians (tuataras) share one form. Varanid lizards and pythons show variants that are partially divided or functionally divided; that may be viewed as a second pattern. These two patterns are those in which the ventricle has three compartments that are incompletely separated from one another: cavum venosum, cavum pulmonale, and cavum arteriosum.[5,6,9-11] A third pattern of a fully divided ventricle characterizes crocodilians.[5,9,12] More recent comparative anatomic work illustrates the details of the cardiac muscle organization, valves, and functional implications in squamates, and a variety of lower to higher vertebrates with particular emphasis on reptiles.[13-15]

The first chamber receiving blood from the system in all reptiles is the sinus venosus—a thin sac-like structure that is often ignored when chambers are described and counted.[11] It can be a large chamber, although it is usually smaller than the combined volume of the atria and ventricle in amphibians, chelonians, and squamates; it is more compact in crocodilians. The sinus venosus is located dorsal to the atria and ventricle.[6] The walls are formed of cardiac muscle and connective tissues. Venous blood from the body drains into the sinus venosus from four major veins: the left precava (left superior vena cava), the right precava (superior vena cava), the left hepatic vein, and the postcava (posterior vena cava). It is connected to the muscular right atrium via a sinoatrial aperture. The sinus venosus is undivided in chelonians and tuataras; a partial septum divides it in squamates and in crocodilians.[10,13]

As in the amphibian, the major aortic vessels are derivatives of the aortic arches. There is some asymmetry in that the pulmonary trunk arises from a single vessel that bifurcates to form the left and right pulmonary arteries. Grossly, it arises alongside the right and left systemic arch from the ventricle. The right aorta gives rise to the carotid arteries as well as branches supplying the body. The left aorta supplies the viscera. The subclavian arteries may arise from the left and right aorta or from dorsal aorta.[5,10,16-18,20]

Venous flow in turtles and lizards is similar to that described for amphibians; however, chelonians have paired abdominal veins, and squamates have a single abdominal vein in most cases. In crocodilians, flow from the hindlimbs often enters the caudal vein or renal portal veins sufficiently caudal to the lateral abdominal veins that flow to the renal postal system may predominate over bypass flow.[20]

Venipuncture Sites

Venipuncture sites in both classes are diverse. Each has its benefits, risks, and biases. In anurans, maxillary (facial), femoral, subclavian, and abdominal veins, as well as cardiac puncture are used; the proximal caudal vein is available in salamanders. Reptilian venipuncture sites include proximal caudal, external jugular, subcarapacial, and brachial veins. Subocular sinus and palatal veins are options in some snakes and lizards, as is cardiac puncture. Consideration of sites selected may reflect potential for differing hematological results, potential for skin discoloration, and risk of accidental puncture of spinal canal. The audience will be requested to discuss their preferred and least favored venipuncture sites.

References

1. Porro LB, Richards CT. Digital dissection of the model organism, *Xenopus laevis* using contrast-enhanced computed tomography. *J Anat.* 2017;231:169–191.

2. Underhill RA. *Laboratory Anatomy of the Frog.* 5th ed. New York: William C Brown Pub.; 1987.

3. Chiasson R. *Laboratory Anatomy of Necturus (Mud Puppy).* New York: William C Brown Pub.; 1976.

4. Gilbert SG. *Pictorial Anatomy of the Necturus.* Seattle, WA: University of Washington Press; 1973.

5. Kardong KV. Vertebrates. In: *Comparative Anatomy, Function, Evolution.* 8th ed. McGraw Hill Higher Ed.; 2018.

6. Farrell AP, Graperil AK, Frances ETB. Comparative aspects of heart morphology. In: Gans C, Gaunt AS, eds. *Biology of the Reptilia.* Society for the Study of Amphibians and Reptiles; 1998:375–424.

7. Kik MJL, Mitchell MA. Reptile cardiology: a review of anatomy and physiology, diagnostic approaches, and clinical disease. *Seminars in Avian and Exotic Pet Medicine.* 2005;14(1):52–60.

8. Seymour R. Scaling of cardiovascular physiology in snakes. *Integr Comp Biol.* 1987;27:97–109.

9. Webb GJW, Heatwole H, de Bavay J. Comparative cardiac anatomy of the reptilia: II. A critique of the literature on the Squamata and Rhynchocephalia. *J Morphol.* 1974;142:1–20.

10. Wyneken J. The structure of cardiopulmonary systems of turtles: implications for behavior. In: Wyneken J, Godfrey M, Bels V, eds. *The Biology of Turtles.* Boca Raton: CRC Press; 2008:213–224.

11. Webb G, Heatwole H, de Bavay J. Comparative cardiac anatomy of the reptilia. I. The chambers and septa of the varanid ventricle. *J Morphol.* 1971;134:335–350.

12. Webb GJW. Comparative cardiac anatomy of the reptilia. III. The heart of crocodilians and an hypothesis on the completion of the interventricular septum of crocodilians and birds. *J Morphol.* 1979;161:221–240.

13. Jensen B, Moorman AF, Wang T. Structure and function of the hearts of lizards and snakes. *Biol Rev.* 2014;89(2):302–336.

14. Jensen B, Wang T, Christoffels VM, Moorman AF. Evolution and development of the building plan of the vertebrate heart. *Biochim Biophys Acta Mol Cell Res.* 2013;1833(4):783–794.

15. Jensen B, Nyengaard JR, Pedersen M, Wang T. Anatomy of the python heart. *Anat Sci Int.* 2010;85(4):194–203.

16. Heisler N, Neumann P, Maloiy GMO. The mechanism of intracardiac shunting in the lizard *Varanus exanthematicus. J Exp Biol.* 1983;105:15–31.

17. Wang T, Altimiras J, Axelsson M. Intracardiac flow separation in an *in situ* perfused heart of Burmese python *Python molurus. J Exp Biol.* 2002;205:2715–2723.

18. Wang T, Altimiras J, Klein W, Axelsson M. Ventricular haemodynamics in *Python molurus*: separation of pulmonary and systemic pressures. *J Exp Biol.* 2003;206:4241–4245.

19. Burggren WW. Form and function in reptilian circulations. *Am Zool.* 1987;27(1):5–19.

20. Holz PH. The reptilian renal portal system—a review. *Bull ARAV.* 1999;9(1):4–14.

Additional references are available from the author (jwyneken@fau.edu).

Speaker Information
(click the speaker's name to view other papers and abstracts submitted by this speaker)

Jeanette Wyneken, PhD
Department of Biological Sciences
Florida Atlantic University
Boca Raton, FL, USA

URL: https://www.vin.com/doc/?id=9768908

Diagnosis and Treatment of the Cardiovascular and Lymphatic Systems of Reptiles and Amphibians

ExoticsCon Virtual 2020 Proceedings

J. Jill Heatley, DVM, MS, DABVP, DACZM

Department of Small Animal Clinical Sciences, College of Veterinary Medicine and Biomedical Sciences, Texas A&M University, College Station, TX, USA

Abstract

Diagnosis and treatment of cardiovascular disease of reptiles and amphibians remain challenging based on lack of normative data, the wide range of adaptive physiologies of these creatures, few case reports, and limited reptile and amphibian specific pharmacokinetic data. Herein I provide an overview of diagnostic and therapeutic approaches for differentiation of reptile and amphibian cardiovascular normality as well as disease and suggestions for treatment.

Introduction

Diagnosis and treatment of cardiovascular disease in reptiles and amphibians remain challenging based on few reported cases, normal baseline data, and fewer successful outcomes.[1-6] Clinical signs may be nonspecific, but those reported will be summarized. Appropriate physical examination and diagnostic approaches will be reviewed as well as cardiovascular treatments. Normal physiological adaptations of the cardiovascular system in reptiles and amphibians, which may obscure diagnosis, will also be highlighted. Therapeutic cardiovascular options will also be summarized. Other than the neoplastic diagnosis of lymphoma, the lymphatic system of reptiles and amphibians appears rarely affected as a primary disease. However, cases of lymphatic disease and expected clinical signs and treatment options will be included.

Cardiovascular and Lymphatic Systems

The cardiovascular and lymphatic systems of reptiles and amphibians will be reviewed in detail by another author; however, for the purposes of this presentation, they will include the heart, veins, arteries, capillaries, and the lymphatic hearts and vessels. These organs serve important roles in delivery of oxygen and other nutrients to organs and cells, maintaining local and systemic fluid homeostasis, and provision of both local and system immune responses. If these organs aren't working, the body's ability to maintain immune, respiration, and fluid homeostasis is, to say the least, severely compromised and death may be imminent.

In general, the reptile and amphibian heart is relatively smaller and works much less than that of most endotherms. In addition, reptile and amphibian innate physiologic tolerance of low blood oxygen and low metabolic rate may obscure diagnosis of cardiovascular abnormalities until end-stage disease.[1] Causes of congestive heart failure in reptiles include restrictive cardiomyopathy, cardiac infection, myocardial fibrosis, atherosclerosis, valvular insufficiency, valvular stenosis, and pericardial effusion.[7] Causes of pericardial effusion in reptiles include infectious pericarditis, noninfectious pericarditis (visceral gout), congestive heart failure, hemopericardium (trauma, aneurysmal rupture, coagulopathy), and neoplasia. Atherosclerosis has been reported in both reptiles and amphibians.

Clinical Signs of Cardiovascular Disease

Often nonspecific in reptiles and amphibians, cardiovascular clinical signs in addition to normal physiological adaptive change vary based on the species and cardiovascular physiology and anatomy, further obscuring the clinical diagnosis (Table 1). Hydration and temperature significantly affect cardiovascular function of both reptiles and amphibians; therefore, the physical examination and diagnostics of the cardiovascular system should be performed while the animal is hydrated and body temperature is within POTZ. For example, in a study of the marine toad (*Bufo marinus*) and the leopard frog (*Rana catesbeiana*), both aerobic and cardiovascular capacity were compromised by dehydration, but were less so in the toad.[8,9] Decline in cardiovascular output/blood flow was mainly based on pulse volume but also due to decreased heart rate. Peripheral resistance increased in both species during dehydration, based on hemoconcentration, but was more profound in the frog based on additional peripheral vasoconstriction likely based on the need to maintain hydration.[8,9] Loss of plasma volume was the principal driver of reduced cardiac output in both of these amphibian species. Evaluation of heart rate in reptiles and amphibians is further made a diagnostic challenge based on the ability of the reptile and amphibian heart rate to readily adapt to the environment (Table 2).[9-11] Herptile heart rate is generally lower than that of similarly sized endotherms. Embryonically, species heart rate varies based on species (at the same temperature, lizards and turtles increased compared to snake and crocodilians) and increased temperature. Embryonic heart rates were also relatively increased in species with smaller adults, smaller eggs, and shorter incubation periods. The total number of embryonic heartbeats between oviposition and hatching was relatively decreased in squamates compared to turtles and crocodilia.[11] Reptile cardiovascular function is best evaluated at POTZ and may be estimated by the following calculation (although this formula overestimates heart rate in bearded dragons.):[2]

Reptile heart rate $_{POTZ}$: $33.4(Wt_{kg}^{-0.25})$

Table 1. Clinical signs of reptiles suffering cardiovascular disease[1-6]

Clinical sign category	Specific clinical signs
Nonspecific	Anorexia, lethargy, depression, inactivity, weight loss, poor BCS, exercise intolerance, sudden death
Cardiac specific	Pericardial effusion, tachycardia, cardiomegaly
Hydrostatic consequences	Edema (cutaneous and renal, hepatic [hepatomegaly], intestinal), pulmonary (pulmonary crackles), ascites
Poor perfusion/oxygenation consequences	Mucosal pallor, cyanosis, open-mouth breathing (reptiles are obligate nasal breathers)

Table 2. Causes of physiologic heart rate change in reptiles and amphibians[1,8,9]

Factor	Increased heart rate	Decreased heart rate
Temperature	Increased ambient temperature	Decreased ambient temperature
Size	Blood loss (surgery, other trauma)	Increased body size

Hemodynamic equilibrium	Trauma, blood loss, handling	Decreased metabolic rate, dehydration, hemoconcentration
Respiration, ventilation, oxygen saturation of blood	Decreased stroke volume	Dive response, breath-holding, decreased respiratory rate
Body sensory stimuli	Basking, vasodilation—maximize heat gain	Cooling, vasoconstriction—minimize heat loss
Positional	Postural and gravitational stress	Postural and gravitational stress

Select cardiac diagnostic approaches for reptiles and amphibians are provided in Table 3. Many basic approaches easily used for companion mammals are challenging to apply in reptiles. Challenging diagnostics include auscultation, which may or may not be cardiac productive dependent upon the species; ECG, which may have low amplitude and few normal values available; peripheral blood pressure reading, which appears completely unrelated to central vascular pressures; ultrasound based on the small windows and heart of many species; and even CT and MRI, based on the challenge of providing contrast and the small heart size. The author finds the usefulness of ultrasound as a diagnostic highly dependent upon the skill of the operator as well as his or her ability to accept the challenge of reptile cardiac ultrasound.

Table 3. Cardiac diagnostic direction for reptile and amphibians[2,4,12,15,16]

Species	Heart position	Doppler crystal placement	Ultrasound probe placement	ECG lead placement
Chelonian	Ventral midline bordered dorsally by lungs, laterally by liver lobes, ventrally by acromion and coracoid processes. Humeral and pectoral scute intersection of ventral midline (right of midline for *Trionychidae* spp.)	Base of the neck, directed caudally and medially	Cervicobrachial window. Direct caudally and medially	Cranial lead: cervical or axillary skin folds, lateral to the neck and medial to the forelimbs. Caudal and neutral lead skin fold caudal to hindlimbs
Lizard* (*Varanids*)	Midline in pectoral girdle (caudal to sternum)	Axillary region, direct medially (lateral/ventral body wall)	Axilla, direct medially (ventral, direct dorsally and cranially)	Cranial lead: cervical region (axilla skin or forelimb) Caudal lead: crural or popliteal fold
Snake[15]	14–45% of total body length. More cranial in arboreal, more caudal in aquatic species	Ventral over heart	Ventral over heart	Base apex reading is taken with electrodes placed 2 lengths cranial/caudal to heart, lateral aspect
Anurans[16]	Underneath sternum, ventral	Ventral over heart	Ventral over heart	RA/LA in ventral axillary regions, LF/RF placed in ventral inguinal regions
Caecilian[12]	Heart position located by palpation	Ventral over heart	Ventral over heart	Electrodes subcutaneously, equilateral triangle of the heart. Standard bipolar and augmented unipolar limb leads were recorded, as was a single precordial lead position

Lizards*—Green iguanas, skinks, chameleons, water dragons

Imaging

Heart size may be difficult to measure on radiographs, based on local tissue confluence and lack of cardiac fat.[1,2,6] Ultrasound, CT, or MRI is recommended to best assess cardiac size, although few normal values are yet available. Creation of vertebral heart score for reptile species in which the heart can be radiographically or CT visualized is recommended based on the following vertebrae as baseline:

- Chelonia—Length/width of the last cervical vertebra unattached to the shell.
- Snakes—The 1st vertebra located cranial to the heart, unaffected by boney change as with previous osteomyelitis, within the lung field for ease of measurement.
- Lizards—The first vertebra completely caudal to the scapular wing. Review of a normal animal for comparison imaging is recommended.

Clinical Signs of Lymphatic Disease

Clinical signs of lymphatic disease are poorly documented in reptiles, but are more prominently reported in amphibians.[3,12,13] This dearth of reptile lymphatic disease reports may be based on relative differences in elasticity and permeability of reptile and amphibian skin.[14] In the amphibian, lymph hearts beat synchronously, about once per second, independent of the heart rate.[14] Certainly, in cases of sepsis, shock, hyperthermia or hypothermia, or vasculitis which is widespread, both systems are likely affected. In the amphibian, suspect illness whenever large volumes of lymph accumulate in any lymph sac. For both species, clinical signs may be expected to be like those of humans suffering lymphatic vascular disease, of which the characteristic clinical sign is lymphedema, which should be cytologically and biochemically differentiated from edema or other fluids. Likely because amphibians have an astounding ability to move lymph rapidly in the normal state (fluid leakage from plasma as high as 50 times the total plasma volume over a 24-hour period), because they have so many lymph hearts (15 in the European fire salamander, *Salamandra salamandra*), and because they have more distensible skin than that of most reptiles, when dysfunction occurs, clinical signs of lymphatic and/or cardiovascular disease in the amphibian may be dramatic and severe.[13] Hydrops (when large amounts of fluid build up in a newborn animal's tissues and organs, causing extreme swelling [reserved for tadpoles]) or anasarca (a general swelling of the whole body that can occur when the tissues of the body retain too much fluid, also known as extreme generalized edema) is commonly reported. Peripheral edema alone may also occur. Signs of dermatosepticemia often accompany these signs but may be difficult to detect based on animal coloration and/or the dilution effect of fluid accumulation.

Diagnosis and Treatment of Lymphatic Disease

In humans, lymphatic vascular insufficiency is often genetic in origin, and diagnosis is confirmed via imaging: indirect radionuclide lymphoscintigraphy, magnetic resonance imaging, and CT. In amphibians and reptiles, I suspect lymphatic disease is more often based on serious systemic disease, as seldom is only a single area clinically affected. Differentials should include cardiac disease, infectious disease (viral and bacterial etiologies are most common), and renal or hepatic disease leading to low total protein and/or vasculitis. Prior to treatment of specific lymphatic disease, the clinician is urged to provide adequate supportive care to achieve POTZ, adequate hydration, proper husbandry to include adequate protein/nutritional intake and antibacterial should other signs of sepsis/vasculitis be present. Therefore, diagnostics should include a CBC, biochemistry, and fecal assessment to rule out common whole-body causes of lymphatic disease: parasitism, other organ failure, and bacterial or viral infection.

Treatment of Cardiovascular Disease

A mini formulary for treatment of cardiovascular disease in reptiles and amphibians is provided (Table 4). Unlike in companion mammals, differentials for reptile and amphibian cardiovascular disease are dominated by infectious and acquired (uric acid or mineralization, atherosclerosis) disease. In many cases, cardiovascular disease may not be manageable by cardiovascular drugs or may be better managed by treatment of underlying husbandry insufficiencies or infectious diseases, to include parasitism. Ensure POTZ and adequate hydration and lack of parasitism. In obese animals, create a nutritional plan for weight reduction and/or specific proteins that may reduce signs of atherosclerosis or uricemia and uric acid deposition. Weakness and lack of mobility due to cardiac insufficiency may limit the animal's ability to obtain food and water, adequate heat, humidity or UVB; rearrangement of the enclosure may be necessary to ensure adequate access. Dietary intake of sodium is usually nonproblematic in reptiles, but overhydration may be problematic in aquatic amphibia; therefore, supplementation with increased sodium and electrolytes either in the enclosure water, drinking or soaking water or food should be considered. Supplementation with fatty acids may be considered, as captive food supplies likely have limited availability of these anti-inflammatory oils compared to free-living counterparts.

Table 4. Cardiovascular drugs for use in reptiles and amphibians[2,5,6,17,18]

Drug	Indication	Species	Dose (mg/kg)
Pimobendan	Calcium sensitizer, positive inotrope and vasodilator, positive lusitrope	Bearded dragon, iguana	0.2 PO q 24 h
Atropine	Parasympatholytic	Reptiles, rarely indicated for bradycardia, appear ineffective in vagal-mediated iguanas	0.01–0.04 IV, IM, SQ
Glycopyrrolate			0.01 IV, IM, SQ
Furosemide	Loop diuretic; may work better in some species than others, despite lack of loop of Henle in these species	Freshwater turtles Iguanas, carpet python, bearded dragon	2,5 2 SQ q 24 h
Hydrochlorothiazide	Thiazide diuretic, calcium-sparing properties, hypertension and fluid retention	Lizard diuretic	1 q 24–72 h
Captopril, enalapril	Less control of arterial blood pressure than phentolamine or phenoxybenzamine	Alligator embryonic experimental use, low long-term dose	5
Lidocaine	Ventricular tachyarrhythmia	Unknown, use low dose	PRN while awaiting other drug effects

References

1. Hynes B, Girling SJ. Cardiovascular and haemopoietic systems. In: *BSAVA Manual of Reptiles.* BSAVA Library; 2019:323–341.

2. Kik MJ, Mitchell MA. Reptile cardiology: a review of anatomy and physiology, diagnostic approaches, and clinical disease. In: *Seminars in Avian and Exotic Pet Medicine.* WB Saunders; 2005;14(1):52–60.

3. Varshney JP. Diagnosis and management of anasarca in a turtle. *Intas Polivet.* 2016;17(1):209–210.

4. Rishniw M, Carmel BP. Atrioventricular valvular insufficiency and congestive heart failure in a carpet python. *Australian Veterinary Journal.* 1999;77(9):580–583.

5. Bogan Jr. JE. Ophidian cardiology—a review. *Journal of Herpetological Medicine and Surgery.* 2017;27(1–2):62–77.

6. Mitchell MA. Reptile cardiology. *Veterinary Clinics of North America: Exotic Animal Practice.* 2009;12(1):65–79.

7. Fitzgerald BC, Dias S, Martorell J. Cardiovascular drugs in avian, small mammal, and reptile medicine. *Veterinary Clinics: Exotic Animal Practice.* 2018;21(2):399–442.

8. Hillman SS. Dehydrational effects on cardiovascular and metabolic capacity in two amphibians. *Physiological Zoology.* 1987;60(5):608–613.

9. Hillman SS, DeGrauw EA, Hoagland T, Hancock T, Withers P. The role of vascular and interstitial compliance and vascular volume in the regulation of blood volume in two species of anuran. *Physiological and Biochemical Zoology.* 2010;83(1):55–67.

10. DeGrauw EA, Hillman SS. General function and endocrine control of the posterior lymph hearts in *Bufo marinus* and *Rana catesbeiana. Physiological and Biochemical Zoology.* 2004;77(4):594–600.

11. Du WG, Ye H, Zhao B, Pizzatto L, Ji X, Shine R. Patterns of interspecific variation in the heart rates of embryonic reptiles. *PloS One.* 2011;6(12).

12. Peters JA, Mullen RK. Electrocardiography in *Caecilia guntheri* (Peters). *Physiological Zoology.* 1966;39(3):193–201.

13. Wright KM, Whitaker BR. Amphibian medicine and captive husbandry. Krieger Publishing Company; 2001.

14. Koltowska K, Betterman KL, Harvey NL, Hogan BM. Getting out and about: the emergence and morphogenesis of the vertebrate lymphatic vasculature. *Development.*

2013;140(9):1857–1870.

15. Bhaskar A, Vinod A. Demonstration of the origin of ECG waves. *Advances in Physiology Education.* 2006;30(3):128.

16. Paillusseau C, Gandar F, Schilliger L, Chetboul V. Two-dimensional echocardiographic measurements in the ball python (*Python regius*). *Journal of Zoo and Wildlife Medicine.* 2020;50(4):976–982.

17. Williams CJ, Alstrup AK, Bertelsen MF, Jensen HM, Leite CA, Wang T. When local anesthesia becomes universal: pronounced systemic effects of subcutaneous lidocaine in bullfrogs (*Lithobates catesbeianus*). *Comparative Biochemistry and Physiology Part A: Molecular and Integrative Physiology.* 2017;209:41–46.

18. Mueller CA, Eme J, Tate KB, Crossley DA. Chronic captopril treatment reveals the role of ANG II in cardiovascular function of embryonic American alligators (*Alligator mississippiensis*). *Journal of Comparative Physiology B.* 2018;188(4):657–669.

Speaker Information
(click the speaker's name to view other papers and abstracts submitted by this speaker)

J. Jill Heatley, DVM, MS, DABVP, DACZM
Department of Small Animal Clinical Sciences
College of Veterinary Medicine and Biomedical Sciences
Texas A&M University
College Station, TX, USA

URL: https://www.vin.com/doc/?id=9768902

Reptile Sedation and Anesthesia

ExoticsCon Virtual 2020 Proceedings

Christoph Mans, DrMedVet, DACZM, DECZM (ZHM)
School of Veterinary Medicine, University of Wisconsin-Madison, Madison, WI, USA

Introduction

Historically, managing anesthesia in reptiles has been problematic. The diversity of reptiles in terms of natural history, size, anatomy, and physiology presents a unique clinical challenge to the veterinarian. Safe and effective anesthesia of the reptilian patient is a requirement for procedures including physical examination of large or dangerous species, collection of diagnostic samples, application of various imaging modalities, as well as for surgical procedures. The anesthetic management of the reptilian patient, including the design of the anesthetic protocol, represents unique challenges due to marked anatomical and physiological differences between taxa. Responses to frequently used anesthetic agents and dosages are often variable between taxa, and individual differences are also commonly observed. Additionally, anesthetic and analgesic protocols (including monitoring techniques to evaluate cardiopulmonary performance) are often derived from domestic animals and then applied to the reptilian patient. These techniques are often of limited value and should be interpreted by clinicians with caution.

Thermoregulation and the Effect of Environmental Temperature

Reptiles are ectothermic, meaning that their body temperature is directly dependent on environmental temperature and associated behaviors to modify that body temperature. Changes in body temperature significantly affect metabolic rate and many other physiologic processes. Therefore, the temperature at which one maintains a patient is an important factor in reptilian anesthesia due to the fact that the absorption, distribution, metabolism, and excretion of drugs in reptiles are directly related to environmental temperature. Induction, duration, and recovery from sedation or anesthesia induced using injectable drugs will be faster if animals are maintained at higher temperatures, and significantly longer if the body temperature is low.[1,2]

Hepatic First-Pass Effect

The hepatic first-pass effect following hindlimb administration of anesthetic and other drugs has not received much attention in the literature. However, its pharmacodynamics and pharmacokinetic significance may be more clinically important than that of the renal portal system.[3] The venous blood flow from the hindlimb in chelonians, lizards, and crocodilians drains into the ventral abdominal vein(s), which either passes directly to the liver or indirectly via the hepatic portal vein.[4] Hence, any liver metabolized or excreted drug administered in the hindlimb first enters the liver before reaching the systemic circulation, resulting in a hepatic first-pass effect. In turtles and crocodiles, the hepatic first-pass effect following hindlimb administration leads to lower bioavailability of drugs, resulting in lower plasma concentrations and/or reduced or no clinical efficacy.[5,6] In leopard geckos, the administration of dexmedetomidine-ketamine in the hindlimbs was significantly less effective in inducing sedation compared to the injection in the forelimbs.[7] In snakes, administration of alfaxalone in the caudal part of the body was significantly less effective in inducing sedation compared to injection in the cranial part of the body.[8] However, despite the published evidence of the clinical significance associated with the hepatic first-pass effect, pharmacokinetic and pharmacodynamic studies continue to be published without taking this anatomical factor into consideration with respect to study design.

Since many drugs commonly used in reptiles, particularly analgesics and anesthetics, undergo hepatic metabolism and/or excretion, one must consider the hepatic first-pass effect, since the majority of venous blood from the hindlimbs and caudal body half drains directly into the liver before entering the systemic circulation. Therefore, it is advised to avoid administration of anesthetic drugs by intramuscular or subcutaneous injection in the caudal body half in reptiles. Intravenous administration of anesthetics in the coccygeal (caudal ventral tail) vein of lizards does not result in a hepatic first-pass effect, because it drains directly into the caudal vena cava. In turtles, the venous drainage from the tail is less consistent, with some blood entering the ventral abdominal vein and some entering the caudal vena cava.

Injectable Anesthetic Agents

Prior to administration of any anesthetic or sedative drugs, reptiles should be maintained within their preferred optimal temperature zone (POTZ), and the core body temperature should be monitored throughout the anesthetic period. A variety of agents representing different classes of drugs are frequently used, either alone or in combination, depending on the desired level of sedation or anesthesia. It is recommended to avoid administration of high dosages of a single anesthetic agent (e.g., ketamine, alfaxalone) and instead consider protocols in which multiple drugs are combined with synergistic actions, thereby requiring lower dosages for each drug. Additionally, using readily reversible drug protocols will provide for more rapid recoveries. Since most deleterious side effects (e.g., prolonged recovery, cardiopulmonary depression) associated with anesthetic and sedative drug administration are dose-dependent, individual drug dosage reduction and reversibility will result in fewer complications and improved recoveries. Similar drugs and drug combinations can also be chosen for premedication/sedation prior to induction of general anesthesia.

If the patient is scheduled for a painful procedure, it is recommended to include an analgesic agent such as an opioid in the preanesthetic protocol. Many injectable protocols in reptiles include the administration of an alpha-2 agonist (e.g., [dex]medetomidine) with ketamine and/or midazolam. A mu-opioid receptor agonist (e.g., morphine, hydromorphone) should be added to the protocol if additional analgesia is required in chelonians and lizards. One must keep in mind that many factors (e.g., site of injection, underlying disease, dehydration, body temperature, gravidity, etc.) may affect the onset, recovery from, and efficacy of anesthetic drugs. Therefore, an individualized approach to choosing appropriate drug combinations and dosages should be considered.

Injectable agents alone can be used to facilitate a variety of short diagnostic and surgical procedures, as well as for induction of general anesthesia. Maintenance of a surgical plane of anesthesia for prolonged, invasive procedures is frequently based on administration of an inhalational agent, such as isoflurane or sevoflurane. In aquatic freshwater turtles and terrestrial chelonians, injectable protocols containing (dex)medetomidine-ketamine, with or without opiates and local anesthetics, were successfully used to induce and maintain surgical anesthesia without the need of gas anesthesia, and these protocols are preferred by the authors.

Routes of Parenteral Drug Administration

While historically the **intramuscular route** of drug administration is recommended in reptiles, it is considered more painful, and less space is provided in small muscle bellies for larger drug volumes when compared to the **subcutaneous route**. In snakes, the epaxial muscles are used, while in lizards and chelonians, muscles of the front limbs are a common intramuscular injection site but provide only limited space to administer anesthetic drugs. The subcutaneous administration of anesthetic and analgesic drugs offers several advantages over intramuscular injection, including access to a variety of subcutaneous sites requiring minimal manual restraint or patient manipulation, suspected less discomfort on injection, and the capability of administering larger volumes in a single location.

The **intravenous route** offers an excellent route of administration for anesthetic agents and results in rapid onset of anesthesia. However, depending on the species and patient size and cooperation, intravenous access may be challenging or impossible to obtain, making the subcutaneous or intramuscular route more feasible. In lizards, intravenous injections are most commonly administered into the caudal (ventral tail) vein. In most tortoises and non-marine turtles, the jugular vein is preferred for intravenous drug administration, but the brachial plexus is also practical. The subcarapacial sinus and dorsal coccygeal vein cannot be recommended for intravenous drug administration in chelonians, due to the risk of accidental administration of anesthetic agents into the intrathecal space, which may lead to severe neurologic complications including permanent paralysis and death. In addition, the dorsal coccygeal vein is not consistently developed across chelonian species, and while present in red-eared sliders, it is absent or poorly developed in several tortoise species. Therefore, it should be assumed that a well-developed dorsal coccygeal vein is the exception, rather than the rule, amongst chelonians. Hence, the author advises against drug administration into the subcarapacial sinus or the dorsal tail vein in chelonians. Intravenous injection sites in snakes are limited to the caudal (ventral tail) vein and the jugular vein following a cut-down procedure. Intracardiac administration of propofol has also been reported in snakes and results in rapid onset with no significant complications reported in a controlled study.[9] However, cardiac tamponade after cardiocentesis in snakes has been reported, and the increased risk of intracardiac injections should be considered.[10]

Drugs

Ketamine, a dissociative agent with dose-dependent anesthetic, sedative, and analgesic properties, is frequently used in reptile anesthesia. However, when administered alone, muscle relaxation is considered inadequate and recoveries excessively prolonged, especially when high dosages (>50 mg/kg) are used. Administration of an alpha-2 adrenergic agonist (e.g., ([dex]medetomidine) and/or a benzodiazepine (e.g., midazolam), in conjunction with ketamine, allows for reduction of ketamine dose. Additionally, these protocols have the benefit of partial reversibility, leading to more rapid recoveries and increased safety. Even at lower, subanesthetic dosages (<5 mg/kg), ketamine can provide additional sedation and analgesia if combined with other anesthetic drugs.

Alpha-2 adrenergic agonists, such as medetomidine and (dex)medetomidine, provide sedation, muscle relaxation, and analgesia in reptiles. Dose-dependent cardiovascular depression was documented in reptiles after alpha-2 agonist administration. Alpha-2 adrenergic agonists are commonly used in combination with ketamine (especially in chelonians) for safe, reliable, and reversible anesthesia but can also be combined with benzodiazepines for procedural sedation. Combining ketamine with medetomidine, or (dex)medetomidine, allows reduction of both drug dosages, and reversibility of the alpha-2 adrenergic agonists with atipamezole will lead to faster and more predictable recoveries. Morphine or hydromorphone can be added to alpha-2 adrenergic agonist plus ketamine protocols in cases in which analgesia is needed. If morphine is used, reversal with naloxone (0.04–0.2 mg/kg SC) should be considered if respiratory depression persists after termination of the anesthetic procedure or recovery is delayed. Concentrated formulations of medetomidine (e.g., 20 mg/ml, ZooPharm Inc.) are available and are more suitable for giant species, such as large chelonians or crocodilians.

Benzodiazepines, such as midazolam and diazepam, have sedative and muscle relaxant properties. Midazolam is water-soluble; can be administered SC, IM, and IV; and is considered a more appropriate choice than diazepam, which is not recommended for IM or SC injection. Midazolam, used alone, is anxiolytic and provides mild, but highly variable sedation, which may be sufficient for minor clinical procedures. Midazolam, combined with ketamine and/or an alpha-2 adrenergic agonist, reduces all drug dosages and attenuates the dose-dependent cardiovascular depressant effects and prolonged recoveries commonly observed with high dosages of ketamine. The effects of benzodiazepines can be antagonized using flumazenil (0.05 mg/kg SC, IM), which shortens recovery from sedation or anesthesia. Concentrated formulations of midazolam (e.g., 50 mg/ml, ZooPharm Inc.) are available and are more suitable for administration of midazolam in giant reptile species.

Propofol, a short-acting, non-barbiturate anesthetic agent can be administered intravenously or intraosseously to provide short-time anesthesia, facilitating procedures such as placement of an esophagostomy tube or abscess debridement, or to induce general anesthesia and allow for endotracheal intubation and maintenance of anesthesia with an inhalational agent. Respiratory depression is also more profound when propofol is administered rapidly. Since propofol does not accumulate in tissues and is rapidly metabolized, recovery can be expected with assisted ventilation in cases of overdose. Propofol requires intravascular administration, which can be challenging. Complications secondary to accidental extravascular injection of propofol intrathecally and into the subcarapacial sinus or the coccygeal vein of chelonian species have been reported.[11,12] Complications associated with intrathecal propofol injection included forelimb and hindlimb paralysis, coma, and spinal necrosis.

Alfaxalone is a short-acting steroid anesthetic, which is labeled for induction of anesthesia in dogs by intravenous administration. However, alfaxalone can be administered by either the intramuscular or subcutaneous route at higher doses. Alfaxalone is rapidly cleared, and its metabolism is independent of organ function; however, pronounced temperature-dependent differences in anesthetic induction, plateau, and recovery times have been reported.[1] Similar to propofol, alfaxalone administration is associated with dose-dependent cardiovascular and respiratory depression in mammals. Recovery from alfaxalone-induced anesthesia is dose dependent, and at high dosages, prolonged recoveries are to be expected. The major advantage of alfaxalone over propofol is that it can be administered intramuscularly as well as subcutaneously, in addition to the intravenous route. Due to the high dosages needed for non-intravenous administration in chelonians and lizards, the resulting drug volumes are usually inappropriately large due to the low concentration of the available alfaxalone formulation (10 mg/ml). Therefore, the author recommends SC injection if large volumes of alfaxalone are to be administered, since the subcutaneous administration of alfaxalone in reptiles has been associated with rapid inductions times.[8,13,14]

Inhalational agents, such as isoflurane and sevoflurane, can be used for both induction and maintenance of anesthesia. Both agents offer the advantages of relatively fast induction and recovery times, as well as limited organ toxicity, especially in patients with renal or hepatic impairment. The low blood solubility allows for rapid changes in anesthetic depth. Administration of inhalational agents follows similar principles as those established for domestic animal anesthesia. Following induction of anesthesia, the trachea should be intubated with an appropriately sized endotracheal tube, which should be connected to a non-rebreathing (body weight <10 kg) or rebreathing system. Intermittent positive pressure ventilation should be performed in all reptiles since they lack a diaphragm and therefore will be apneic while under a surgical plane of anesthesia. The concentration of the inhalational agent necessary to maintain a surgical plane of anesthesia depends on the health status of the patient. Minimum anesthetic concentrations (MAC) of isoflurane and sevoflurane were determined in green iguanas and are 1.8–2.1% and 2.1–4.1%, respectively.[15] Maintenance requirements for a surgical plane of anesthesia were reported to be 2–3% for isoflurane and 3.5–4.5% for sevoflurane. Both isoflurane and sevoflurane have a dose-dependent, depressive effect on the cardiovascular system. Isoflurane was shown to significantly reduce blood pressure in green iguanas and ball pythons. Therefore, significant cardiovascular depression can occur with normal maintenance concentrations of isoflurane. No significant differences were detected between sevoflurane and isoflurane in cardiopulmonary function of green iguanas. However, use of sevoflurane resulted in faster induction and recovery compared to isoflurane.

Due to right-to-left cardiac shunting in some reptiles, reduced lung perfusion can occur, and therefore, concentrations of inhalant anesthetic gases in the lungs do not necessarily reflect the concentrations in the blood or brain. Sudden changes in shunting directions can lead to sudden changes in inhalant anesthetic blood concentration, which can lead to significant changes in anesthetic depth, evidenced by slow induction, sudden arousal from anesthesia, inability to deepen a lightly anesthetized reptile, or prolonged recovery from anesthesia. In lizards and snakes, changes in gas concentration can be used to control anesthetic depth, but in chelonians—particularly aquatic turtles—a change in gas concentration will not lead to measurable changes in anesthetic depth as a function of cardiac shunting, breath-holding ability, and ventilation-perfusion mismatch. Intracardiac shunting can be eliminated in chelonians by administration of atropine, which results in lower MAC of isoflurane.[16]

General Anesthesia

Induction

Induction of anesthesia can be accomplished with injectable or inhalational agents. In patients with established, reliable vascular access, slow administration of propofol (3–10 mg/kg IV, IO) or alfaxalone (5–10 mg/kg IV) will induce rapid induction of anesthesia. Alternatively, intramuscular or subcutaneous administration of various combinations of ketamine, midazolam, alpha-2 agents, alfaxalone, and opioids will induce a dose-dependent plane of anesthesia adequate for endotracheal intubation and maintenance with an inhalational agent, if required. Mask or chamber induction with an inhalational agent can be used in snakes and lizards, and for handler safety reasons is particularly useful in venomous species. Mask or chamber inductions require high concentrations of gas anesthetic, which is associated with environmental contamination and can contribute to unpredictable and prolonged induction times. In monitor lizards, induction times of >10 minutes were noted following isoflurane and sevoflurane delivery by mask. In reptile species that are able to breath-hold for extended periods (in particular, semiaquatic and aquatic turtles and crocodilians), induction times will be prolonged, and/or induction with inhalant agents may be ineffective.

Endotracheal intubation is relatively easy to perform in most snakes and lizards. In these species, the glottis is located rostrally in the oral cavity, and the trachea can be intubated with an appropriately sized endotracheal tube.

Maintenance

During maintenance of anesthesia, the patient should be monitored regularly for depth of anesthesia, cardiopulmonary performance, and evidence of pain. If indicated, the anesthetic level should be adjusted, and additional analgesic therapy may be warranted. Similar to domestic animals, cardiopulmonary depression is dose dependent in reptiles, and a decrease in heart rate and arterial blood pressure will be observed—although blood pressure monitoring is difficult and frequently impractical.

All reptiles require intermittent positive pressure ventilation (IPPV) at a surgical plane of anesthesia to maintain $EtCO_2$ between 10 and 25 mm Hg. Over-ventilation (decreased $EtCO_2$) causes respiratory alkalosis and delays return to spontaneous ventilation during recovery. Ventilation can be performed manually or more accurately with a mechanical ventilator. Any anesthetized reptile that is breathing spontaneously is light. While anecdotally it has been recommended to ventilate an anesthetized reptile at 4–6 breaths per minute, the rate of IPPV needs to be adjusted to each individual patient based on the preanesthetic examination and based on $EtCO_2$ values. For example, in anesthetized neotropical rattlesnakes (*Crotalus durissus*), 5 breaths per minute led to respiratory alkalosis and was considered excessive, while 0.67 breaths per minute led to mild hypercapnia and mild respiratory acidosis.[17] Therefore, a respiratory rate of 1–2 breaths per minute at a tidal volume of 30 ml/kg was recommended for this species.

Monitoring Techniques

Accurate monitoring of anesthetic parameters is essential to safely and effectively anesthetize the reptilian patient. Application and interpretation of many monitoring techniques in reptiles are in their infancy, and some devices and techniques may only be of limited value. Monitoring of cardiopulmonary performance in the anesthetized reptilian patient presents challenges, and further work is needed to establish reference data for common reptile species. One should be familiar with the applications and limitations of various monitoring devices developed for human and veterinary anesthesia. Anesthesia monitoring techniques such as pulse oximetry, arterial blood gas analysis, and capnography are commonly used in human and veterinary anesthesia, but not all have been validated for use in reptiles. In addition, reference values for common reptile species and common anesthetic protocols are scant. Therefore, knowledge of the anatomy, physiology, and pathophysiology is essential. Also, the pharmacology of sedative and anesthetic agents being used, as well as their effects on cardiopulmonary performance and organ function, must be known. In order to accurately interpret cardiopulmonary performance, there is a need for normal values in conscious reptiles; however, few studies have investigated normal cardiopulmonary parameters in conscious, non-anesthetized reptiles.

Reflexes—Determination of the degree of sedation and depth of anesthesia is commonly performed by assessment of the presence or absence of reflexes, such as head or limb withdrawal, righting reflex, palpebral and corneal reflexes, toe/tail pinch reflex, as well as cloacal tone. However, significant anatomic differences exist between snakes, chelonians, and lizards. The corneal and palpebral reflex is present and useful in many reptiles but cannot be assessed in snakes and most geckos (one exception being leopard geckos), due to lack of eyelids and presence of spectacles. The righting reflex should be absent in snakes and lizards during a surgical plane of anesthesia, but it is not as useful in turtles and tortoises. Assessment of head withdrawal and neck tone may provide useful information in turtles and tortoises. Snakes tend to lose muscle tone from head to tail during induction and regain muscle tone from tail to head during recovery. In general, a surgical plane of anesthesia, including effective analgesia, is determined when there is no response to a painful stimulus (such as a skin incision), and consequently, no changes in cardiopulmonary parameters such as tachycardia, hypertension, or an increase in respiratory rate.

Recovery

Following cessation of the procedure, inhalant anesthesia should be discontinued, and any reversal drugs administered (taking care to provide alternative analgesia if using naloxone). Room air administered with an Ambu bag™ instead of 100% oxygen has, historically, been recommended for ventilation of reptiles during recovery. It was suggested that high oxygen concentrations in the lungs delay return to spontaneous respiration. However, recent studies could not demonstrate any statistical or clinically significant difference in recovery times in monitor lizards or bearded dragons ventilated with 100% oxygen, compared to animals ventilated with room air during recovery from gas anesthesia.[18]

As is standard protocol for domestic animals, fluid therapy and pain management should be continued during the recovery period, and additional analgesia should be provided as necessary. Prolonged anesthetic recoveries are common in reptiles, especially if high dosages of non-reversible injectable drugs are administered (e.g., tiletamine/zolazepam). Isoflurane anesthesia in chelonians is also frequently associated with slow recoveries, suspected to be due to the hypotension, right to left shunting, and reduced lung and tissue perfusion. Administration of epinephrine (syn. adrenalin, 0.1 mg/kg IM) in common snapping turtles (*Chelydra serpentina*) significantly reduced time (>50%) to return of spontaneous breathing and complete recovery following isoflurane anesthesia. It was demonstrated that an increase in adrenergic tone of the pulmonary vasculature leads to shunting of blood away from the pulmonary circulation during episodes of apnea in turtles, and that by increasing the sympathomimetic tone of the systemic vasculature through administration of epinephrine, pulmonary shunting is reduced, leading to accelerated recovery in turtles anesthetized with isoflurane.[19] Stimulation of the acupuncture point GV-26 also hastens snapping turtle recovery from isoflurane, comparable to the administration of epinephrine.[19]

During recovery, the patient should be kept in a temperature-controlled environment. For most reptilian patients, dedicated reptile or small animal incubators are ideal since they are designed for accurate temperature maintenance. It is not recommended to exceed the species-specific POTZ since this will result in an increased metabolic rate and oxygen demand. Reptiles recovering from general anesthesia will be in a state of respiratory compromise, and it may be difficult to meet the increased demand of oxygen by spontaneous respiration. During the recovery period, it is essential to assess respiratory status of the patient and continue assisted ventilation until the patient is breathing spontaneously, has regained reflexes, and can be extubated.

Spontaneous breathing and ambulation are reliable indicators of anesthetic recovery. Reptiles should be extubated once spontaneous breathing is observed and oropharyngeal reflexes (e.g., jaw tone, tongue movement) have returned. If spontaneous breathing has not returned, IPPV should be continued. Even following extubation, monitoring should continue, as it is not unheard of for a spontaneously breathing reptile to slip back into unconscious apnea.

Regional Anesthesia and Analgesia

Regional anesthesia and analgesia are currently underutilized in reptile medicine but offer significant benefits; for example, the ability to reduce injectable, inhalant anesthetic, and sedative drug requirements, and to produce preemptive analgesia in the perioperative period without deleterious side effects (such as respiratory depression). In addition, drugs used for regional anesthesia are widely available and do not require controlled drug scheduling. Finally, in mammals, local anesthetics have the potential to blunt a surgery-induced neuroendocrine (stress) response. Knowledge of anatomical landmarks and basic pharmacology of local anesthetics is required for safe and effective nerve blocks. Regional anesthesia can be used as a sole anesthetic modality in manually restrained or sedated reptiles for minor surgical procedures. Common indications for the use of local anesthetics as part of an anesthetic protocol in reptiles include: distal limb surgery, cloacal procedures (e.g., prolapse, cloacoscopy, egg removal), phallectomy, and celiotomy and celioscopy. Local anesthetics can be administered as part of topical or incisional (infiltration) anesthesia, peripheral or cranial nerve blocks, or neuraxial (intrathecal, spinal) anesthesia.

Clinical experience demonstrates that local anesthetics have similar onset and duration of action in reptiles when compared with mammals. Dosages, volumes, and concentrations of local anesthetics are crucial in determining the magnitude and quality of the anesthetic block. High concentrations are preferred, but sufficient volume of tissue distribution should be considered. Site of injection will be an important factor in inducing the onset of block. For shorter onset of anesthesia, local anesthetics should be administered as close as possible to, but not directly into, a nerve. Physical and chemical characteristics of the local anesthetic explain the slow onset, but some lipophilic drugs (e.g., bupivacaine) provide a long duration of action.

Local anesthetics can produce irreversible or reversible nerve damage. Neurotrauma occurs during intraneural injections. A local anesthetic should never be injected if resistance to injection is observed. Systemic toxicity is produced by accidental intravascular injection of local anesthetics. Toxic dosages associated with adverse effects have not been determined in reptiles, but maximum dosages used in mammals—particularly in small reptiles—should be used as a guide. Adverse effects can include seizures, cardiorespiratory depression, and death. Only preservative-free drug formulations should be used for intrathecal injections.

Lidocaine (2%) is an aminoamide local anesthetic that is commonly used in reptile medicine. Dosages associated with toxicity were reported to be between 5 and 20 mg/kg in mammals. In reptiles, lidocaine was administered at up to 4 mg/kg intrathecally without side effects. Concentrations of 0.5% and 1% are also available. In addition, a commercial mixture of lidocaine (2.5%) and prilocaine (2.5%) in gel preparation (EMLA® cream) is available and used for venipuncture in small animals. Its use was reported in turtles and tortoises undergoing phallectomy. Infiltration of the prefemoral fossa in chelonians with lidocaine has been reported and is frequently used by biologists for coelioscopy. However, it has been documented that this technique using 1 mg/kg of lidocaine results in insufficient anesthesia and analgesia compared to general anesthesia, and is therefore not recommended.[20]

Bupivacaine is a highly lipophilic, aminoamide local anesthetic that produces significant cardiotoxicity when injected intravenously. It is not clear what dosages will produce these deleterious effects in different reptile species, but negative aspiration of blood into the needle hub is always recommended prior to bupivacaine administration.

Regional Anesthetic Techniques

Tissue infiltration (incisional anesthesia)—Local anesthetics are used for incisional anesthesia and can be infiltrated in the subcutaneous tissue in association with a surgical field; for example, prior to celiotomy ("line block"), skin biopsies, or laceration repair. However, local anesthetics should not be administered in proximity to neoplastic tumors since the process can spread neoplastic cells. In small animals, this technique is used most frequently before laparotomy and as part of a multimodal analgesic approach, during which NSAIDs are also administered systemically. Aseptic techniques are always employed.

Nerve blocks—Local anesthetic techniques of the head and limbs are not commonly reported in the reptile literature. This is likely due to the large number of reptilian species and the paucity of information on their specific neuroanatomy and the efficacy of locoregional techniques. In crocodilians, dental blocks have been reported.[21]

Spinal anesthesia—Spinal (also referred to as **intrathecal** or **subdural**) anesthesia and analgesia of the spinal cord are currently in their infancy in reptiles and have only been reported in turtles and tortoises. In chelonians, epidural anesthesia is not possible due to the lack of a sufficiently developed epidural space. However, a well-developed intrathecal (subdural) space, which directly surrounds the spinal cord and is filled with cerebrospinal fluid (CSF), has been published and allows for intrathecal administration of various anesthetic and analgesic drugs. The presence of the carapace, and the fusion of the vertebral column to the carapace, limits access only to the cervical and coccygeal intrathecal space in chelonians, but only intrathecal injections in the coccygeal region are of clinical interest.

Spinal anesthesia has been evaluated in several species of chelonians. In red-eared sliders (*Trachemys scripta elegans*), spinal anesthesia was successfully induced in turtles of both sexes, with body weights ranging from 0.5 to 1.0 kg.[22] Injections were performed in sedated turtles, which allowed for proper restraint of the tail and correct intrathecal administration at the level of the mid-to-proximal coccygeal vertebrae. The turtles were positioned in ventral recumbency, and the tails were aseptically prepared before insertion of 28-G needles, attached to 0.5-mL insulin syringes. Successful intrathecal injection and induction of spinal anesthesia were confirmed by complete motor block (relaxation) of the tail, cloacal sphincter, and hindlimbs. The onset to motor block ranged from 1 to 5 minutes. Following the initial injection of either preservative-free lidocaine (4 mg/kg, 2%) or bupivacaine (1 mg/kg, 0.5%), spinal anesthesia was successfully induced in approximately half of the turtles. In turtles showing no evidence of spinal anesthesia after the initial injection, a second intrathecal injection (of the same drug and dosage) was administered approximately 15 minutes following the first. The repeated intrathecal injection increased the overall success rate to approximately 80–90%. In male and female red-eared sliders, the duration of motor block of the tail, cloacal sphincter, and hindlimbs was approximately 1 hour following intrathecal lidocaine administration (4 mg/kg, 2%) and about 2 hours following intrathecal bupivacaine injection (1 mg/kg, 0.5%). There was a large variation in duration of spinal anesthesia in turtles of both genders. A possible explanation for this variability was associated with the well-developed and prominent intravertebral venous plexus, which is located within the spinal canal directly overlying the intrathecal space, both dorsally and laterally.

References

1. Kischinovsky M, Duse A, Wang T, *et al.* Intramuscular administration of alfaxalone in red-eared sliders (*Trachemys scripta elegans*)—effects of dose and body temperature. *Vet Anaesth Analg.* 2013;40:13–20.

2. Olsson A, Phalen D. The effects of decreased body temperature on the onset, duration and action of medetomidine and its antagonist atipamezole in juvenile farmed estuarine crocodiles (*Crocodylus porosus*). *Vet Anaesth Analg.* 2013;40:272–279.

3. Holz PH. The reptilian renal portal system—a review. *Bull Assoc Reptil Amphib Vet.* 1999;9:4–14.

4. Benson KG, Forrest L. Characterization of the renal portal system of the common green iguana (*Iguana iguana*) by digital subtraction imaging. *J Zoo Wildl Med.* 1999;30:235–241.

5. Kummrow MS, Tseng F, Hesse L, *et al.* Pharmacokinetics of buprenorphine after single-dose subcutaneous administration in red-eared sliders (*Trachemys scripta elegans*). *J Zoo Wildl Med.* 2008;39:590–595.

6. Olsson A, Phalen D. Preliminary studies of chemical immobilization of captive juvenile estuarine (*Crocodylus porosus*) and Australian freshwater (*C. johnstoni*) crocodiles with medetomidine and reversal with atipamezole. *Vet Anaesth Analg.* 2012;39:345–356.

7. Fink DM, Doss GA, Sladky KK, *et al.* Effect of injection site on dexmedetomidine-ketamine induced sedation in leopard geckos (*Eublepharis macularius*). *J Am Vet Med Assoc.* 2018;253:1146–1150.

8. Yaw TJ, Mans C, Johnson SM, *et al.* Effect of injection site on alfaxalone-induced sedation in ball pythons (*Python regius*). *J Small Anim Pract.* 2018;59:747–751.

9. McFadden MS, Avery Bennett R, Reavill DR, *et al.* Clinical and histologic effects of intracardiac administration of propofol for induction of anesthesia in ball pythons (*Python regius*). *J Am Vet Med Assoc.* 2011;239:803–807.

10. Selleri P, Di Girolamo N. Cardiac tamponade following cardiocentesis in a cardiopathic boa constrictor imperator (*Boa constrictor imperator*). *J Small Anim Pract.* 2012;53:487.

11. Quesada RJ, Aitken-Palmer C, Conley K, *et al.* Accidental submeningeal injection of propofol in gopher tortoises (*Gopherus polyphemus*). *Vet Rec.* 2010;167:494–495.

12. Innis CJ, DeVoe R, Mylniczenko N, Garner M. A call for additional study on the safety of subcarapacial venipuncture in chelonians. In: *Proc Assoc Rep Amph Vet.* 2010:8–10.

13. Doss GA, Fink DM, Sladky KK, *et al.* Comparison of subcutaneous dexmedetomidine–midazolam versus alfaxalone–midazolam sedation in leopard geckos (*Eublepharis macularius*). *Vet Anaesth Analg.* 2017;44:1175–1183.

14. Ratliff C, Parkinson LAB, Mans C. Effects of the fraction of inspired oxygen on alfaxalone-sedated inland bearded dragons (*Pogona vitticeps*). *Am J Vet Res.* 2019;80:129–134.

15. Barter LS, Hawkins MG, Brosnan RJ, *et al.* Median effective dose of isoflurane, sevoflurane, and desflurane in green iguanas. *Am J Vet Res.* 2006;67:392–397.

16. Bertelsen MF, Hansen K, Williams C, *et al.* Elimination of intracardiac shunting provides stable gas anesthesia in tortoises. *Sci Rep.* 2018;8:1–6.

17. Bertelsen MF, Buchanan R, Jensen HM, *et al.* Assessing the influence of mechanical ventilation on blood gases and blood pressure in rattlesnakes. *Vet Anaesth Analg.* 2015;42:386–393.

18. Odette O, Churgin SM, Sladky KK, *et al.* Anesthetic induction and recovery parameters in bearded dragons (*Pogona vitticeps*): comparison of isoflurane delivered in 100% oxygen versus 21% oxygen. *J Zoo Wildl Med.* 2015;46:534–539.

19. Goe A, Shmalberg J, Gatson B, *et al.* Epinephrine or GV-26 electrical stimulation reduces inhalant anesthetic recovery time in common snapping turtles (*Chelydra serpentina*). *J Zoo Wildl Med.* 2016;47:501–507.

20. Hernandez-Divers SJ, Stahl SJ, Farrell R. An endoscopic method for identifying sex of hatchling Chinese box turtles and comparison of general versus local anesthesia for coelioscopy. *J Am Vet Med Assoc.* 2009;234:800–804.

21. Wellehan JFX, Gunkel CI, Kledzik D, *et al.* Use of a nerve locator to facilitate administration of mandibular nerve blocks in crocodilians. *J Zoo Wildl Med.* 2006;37:405–408.

22. Mans C. Clinical technique: intrathecal drug administration in turtles and tortoises. *J Exot Pet Med.* 2014;23:67–70.

Speaker Information
(click the speaker's name to view other papers and abstracts submitted by this speaker)

Christoph Mans, DrMedVet, DACZM, DECZM (ZHM)
School of Veterinary Medicine
University of Wisconsin-Madison
Madison, WI, USA

The Reptilian and Amphibian Circulatory Systems: The Pathology

ExoticsCon Virtual 2020 Proceedings

Drury Reavill, DVM, DABVP (Avian and Reptile/Amphibian Practice), DACVP; Ethan Biswell, DVM, DACVP
ZNLabs, Salt Lake City, UT, USA and Louisville, KY, USA

Abstract

This paper will cover the diseases of the cardiovascular system in reptiles and amphibians. It is not intended as a comprehensive review. In-depth disease descriptions can be found in the references cited. The paper is divided by etiologic disease categories involving the cardiovascular system. A short discussion of the disease conditions occurring in reptiles and amphibians is found in each section.

Introduction

Poikilotherms have been considered to have few primary cardiovascular disorders.[1] Clinical cardiovascular disorders may be primary or secondary and are associated with infection, parasitism, nutritional imbalances, and poor husbandry practices.[2] This section will present a limited review of reptile cardiovascular pathology.

Developmental Anomalies

Developmental anomalies are reported to be common in reptiles[3] and are usually associated with deaths of juveniles.[1] Incomplete development of the atrioventricular valves has been described, leading to incomplete valve closure.[4] A secundum atrial septal defect was described in a Komodo dragon (*Varanus komodoensis*).[5] The lesion was on the craniodorsal portion of the septum. It led to chronic congestion of several organs. Two juvenile ball pythons (*Python regius*) with bifid ventricles were reported.[6] All cardiac chambers were enlarged; the lesion disrupted the ability to separate pressures within the ventricles. An interventricular septal defect was also reported in an alligator (*Alligator mississippiensis*).[7]

Noninfectious Diseases

Aortic aneurysm rupture leading to cardiopulmonary arrest was described in a Burmese python (*Python molurus bivittatus*), and the exact cause was not determined.[1] Dissecting aortic aneurysm in a bearded dragon presented with acute listlessness, weakness, and general pallor. A pericardial hemorrhagic effusion produced an enlarged cardiac silhouette on radiographs.[8] Arteriosclerosis has been documented in green iguanas (*Iguana iguana*),[9] and dystrophic mineralization of arteries is seen in various lizards.[4] Mineralization of arteries is most likely due to secondary nutritional hyperparathyroidism from a calcium deficiency. A mole king snake (*Lampropeltis calligaster rhombomaculata*)[10] and a Deckert's rat snake (*Elaphe obsoleta deckerti*)[11] were diagnosed with cardiomyopathy of undetermined cause. The lesions in the king snake were primarily collagen proliferation and osteoid-like material, while the rat snake had degeneration and necrosis of myocardial fibers. Aortic stenosis with associated dilatation of the right atrium and the ventricle were found in an iguana.[12] Histologically, both aortic arches had narrow lumens with thickened intimae, and the myocardial fibers were attenuated. Although the cause was not determined, the authors considered a chronic congenital lesion as one possibility.

A bearded dragon with severe atherosclerosis was also associated with hypertensive heart disease. The adult animal had bradyarrhythmia and balance abnormalities. Cardiac function was evaluated with electrocardiography (ECG) and echocardiography, which identified an AV block. Atherosclerosis is uncommon in reptiles.[13]

Myocardial degeneration due to vitamin E deficiency has been seen in several reptile species.[14] Grossly and microscopically, the lesion is similar to that described in mammals. The heart will have mottled white to grey myocardium, and histologically there will be multifocal to coalescing loss of myocytes replaced with dense fibrous stroma. Visceral gout can lead to severe pericardial lesions,[15] as well as changes in the myocardium. These can include thickening of the pericardial sac with a deposition of the urate crystals as well as crystals forming within the myocardium. Primary disease can be nutritional or toxic.[16]

Infectious Diseases

Infectious agents affecting the heart are usually secondary to systemic illness.[2] West Nile virus infection led to myocardial degeneration and necrosis in a farmed American alligator.[17] Arenavirus (inclusion body disease of boids) inclusions have been identified in cardiac myocytes; however, they do not appear to be associated with any clinical changes.[18]

Iridovirus has been recognized as systemic infections in both salamanders and frogs. Infections in frogs are characterized by increased tadpole mortality and are best described in the tadpoles of the American bullfrog. There will be hemorrhage identified in the skin and muscle, as well as stomach. Necrotic foci are recognized in the liver, kidney, and intestines. A viral epizootic has been recognized in larval and adult tiger salamanders. The primary lesion was a skin lesion with cutaneous ulcers and erosions. These epithelial cells contain large cytoplasmic inclusions. Systemic effects were noted; however, no specific lesions were associated with the direct viral infection. Intracytoplasmic viral inclusions were in the heart sections.[19,20]

Mycoplasma infection was seen in American alligators (*Alligator mississippiensis*).[21] Rapidly growing mycoplasmas were identified by culture and polymerase chain reaction. The alligator isolate is a novel species in the *Mycoplasma* family.

Bacterial disease has been reported in conjunction with endocardial thrombosis,[22] endocarditis,[23] endocarditis valvularis,[24,25] and myocarditis.[26] A bacterial infection resulting in a necrotizing myocarditis and multiple granulomas in the kidney was suspected in a Boa constrictor. The lesion was suspected to have been due to a bite wound on the ventrum, and that may have penetrated the underlying heart.[27] A severe bacterial pneumonia (*Salmonella enterica* ssp. IV, *Enterobacter cloacae*, and *Klebsiella pneumoniae*) in a Burmese python was associated with endocarditis of the pulmonic valve and a dilated truncus pulmonalis. An additional lesion of arteriosclerosis of the pulmonary trunk was present. Grossly, the lesion on the pulmonic valve was a cauliflower-like mass lesion.[25] Valvular endocarditis is described in an Argentine boa (*Boa constrictor occidentalis*) with a concurrent arenavirus infection. A vegetative lesion was present on the right atrioventricular valve with valvular insufficiency and a mildly dilated right atrium. The granuloma on the cardiac valve contained gram-positive cocci. In addition, there was a subacute bronchopneumonia and air sacculitis with edema.[24] Mycobacterial infections, although uncommon in reptiles, have involved the heart.[28]

Systemic gram-negative bacterial septicemias commonly involve the cardiovascular system in amphibians. The frequent clinical presentation is of generalized ascites or localized edema. The best-known bacteria in these cases include *Aeromonas hydrophila*, *Flavobacterium* spp., *Klebsiella pneumoniae*, *Proteus mirabilis*, and *Chlamydophila* species.[29,30] Mycobacterial infections are common in some species of amphibians. These are also generally systemic infections, involving the heart.[28]

Chlamydophila sp. infection is a cause of heart disease in reptiles. Granulomatous pericarditis and myocarditis were found in puff adders (*Bitis arietans*)[31] and necrotizing myocarditis in green turtles (*Chelonia mydas*).[32] Organisms could be demonstrated in macrophages in sections of paraffin-embedded heart. In emerald tree boas (*Corallus caninus*), there were histiocytic granulomas in the heart as well as in other organs. Small, basophilic organisms were seen on hematoxylin and eosin-stained sections. Transmission electron microscopy of an intestinal granuloma demonstrated developmental stages of the chlamydial organisms.[33] A combined infection of *Chlamydia* species and *Aeromonas sobria* resulted in a systemic infection including the liver, spleen, kidneys, heart, and the respiratory tract in farmed juvenile Siamese crocodiles (*Crocodylus siamensis*). The inflammation was a severe necrotizing lymphohistiocytic and heterophilic response.[34]

Systemic fungal infections commonly involve the heart in various amphibians.[35] The most commonly reported systemic fungus in amphibians is Chromomycosis. The lesion is associated with pigmented fungal organisms from the genera *Cladosporium*, *Fonsecaea*, *Phialophora*, *Rhinocladiella*, *Hormodendrum*, *Curvularia*, and *Drechslera*, among others. These fungi are saprophytes that live in soil and decaying vegetable matter, and they are potentially zoonotic. In tissue, they produce characteristic amber-brown, thick-walled septate structures known as sclerotic bodies. Chromomycosis generally presents first as an ulcerated or granulomatous skin lesion. These lesions can be nodules, grey-black papules, and/or irregularly shaped grey-black ulcers. The most common sites for external lesions are the dorsal and ventral skin surfaces. Over a period of time (which can be >6 months), the animal succumbs to disseminated disease. On gross examination, multiple organs may contain multiple coalescing grey-black nodules. These granulomas are commonly described in the liver and kidney, but they also occur in the spleen, lung, heart, adrenal, gastrointestinal tract (serosal and mucosal surfaces), bone, and central nervous system.[36] Exfoliative cytology of the cutaneous lesions may have mononuclear cells surrounding pigmented, septate fungal hyphae. These organisms are zoonotic with human infection resulting from the contaminated environment.

Encephalitozoon pogonae can result in a systemic infection in bearded dragons. The infection has occurred in the great vessels in the heart, resulting in a granulomatous arteritis. An acute and severe hemopericardium developed in two cases.[37]

Spirorchid fluke (Digenea: Spirorchidae) infections have been recognized in a variety of marine turtles. From necropsies of 96 stranded green turtles, cardiovascular lesions included mural endocarditis, arteritis, and thrombosis—frequently accompanied by aneurysm formation.[38] Spirorchiid trematode eggs were noted in the heart and other organs in black turtles (*Chelonia mydas agassizii*).[39] Seventy-five adult *Learedius learedi* were recovered from the heart of one of the turtles.[39] Loggerhead turtles (*Caretta caretta* L.) supported lesions from two spirorchiids, *Hapalotrema mistroides* and *Neospirorchis* species. The lesions were mild with no obvious clinical signs. Rarely vasculitis was observed in the heart and great vessels.[40]

During a 5-year period, 16 freshwater turtles (*Trachemys scripta elegans* and *Chrysemys picta*) died spontaneously.[41] Necropsy lesions included subcutaneous edema, hepatic necrosis, pancreatic necrosis, splenic necrosis, and intestinal parasites. Histologically, trematode eggs were seen within the myocardium and in other tissues, associated with granulomatous lesions. The size and distribution of the eggs were consistent with *Spirorchis* sp. infection. Fluke infections can result in parasites within the three chambers of the heart and in major vessels (right aortic arch and brachiocephalic artery), attached to the walls, or free in the lumen.[42]

Whiptail lizards (*Cnemidophorus* spp.) from Texas and Colorado (USA) were examined for *Mesocestoides* sp. tetrathyridia, and 11 (5%) were infected. Free tetrathyridia were found in the body cavity of lizards, and encapsulated tetrathyridia were observed in the heart, liver, and stomach.[43]

The invasive Coqui frogs (*Eleutherodactylus coqui*), as well as Cuban greenhouse frogs (*Eleutherodactylus planirostris*) and cane toads (*Rhinella marina*) serve as paratenic hosts for the lungworm (*Angiostrongylus cantonensis*), an important human health concern in Hawaii. Larval migration was identified through the stomach, intestines, muscle, liver, heart, and brain.[44]

Neoplastic Diseases

Primary tumors of the reptile cardiovascular system are not common. Among those reported are a rhabdomyosarcoma in a boa constrictor (*Boa constrictor*), a hemangiosarcoma in the heart of a Madagascar giant hognose snake (*Leioheterodon madagascariensis*; family Colubridae),[45] and a fibrosarcoma in a Gaboon viper (*Bitis gabonica*).[46] Other tumors that may be primary or part of a disseminated process include lymphoblastic malignant lymphoma in several species of reptiles[46] and malignant chromatophoromas in bearded dragons.[47] Thirty-nine percent of green turtles with fibropapillomatosis (FP) had disseminated internal tumors, most of them in the lung, kidney, and heart.[48] A disseminated mast cell tumor diagnosed in an eastern kingsnake (*Lampropeltis getulus getulus*) involved the heart and numerous other organs.[49] A disseminated chondrosarcoma involved the heart in a corn snake (*Elaphe guttata*).[50] Large areas of myocardium had been replaced by neoplastic tissue. A metastatic oviductal adenocarcinoma has also been reported in the heart of a corn snake.[51]

Vascular tumors include hemangiosarcomas and hemangiomas. These are uncommon tumors in reptiles. A corn snake developed a mid-body swelling, and the extraluminal mass was a splenic hemangiosarcoma.[52] A similar case has been reported in a lizard.[53] Hemangiomas are also very rare, with two cases reported: a corn snake with an atrial cardiac tumor and an esophageal tumor in a red-eared slider.[54,55] Some tumors that have been tested were positive for factor VIII-related antigen, by immunohistochemistry.[45]

Spontaneous and primary cardiac tumors of amphibians are not reported in the literature (VIN and PubMed search, April 2020).

References

1. Rush EM, Donnelly TM, Walberg J. What's your diagnosis? Cardiopulmonary arrest in a Burmese python. Aortic aneurysm. *Lab Anim (NY)*. 2001;30(3):24–27.

2. Murray MJ. Cardiology. In: Mader DR, ed. *Reptile Medicine and Surgery*. Philadelphia, PA: WB Saunders; 2006:181–195.

3. Mitchell MA. Reptile cardiology. *Vet Clin Exot Anim*. 2009;12:65–79.

4. Kik MJL, Mitchell MA. Reptile cardiology: a review of anatomy and physiology, diagnostic approaches, and clinical disease. *Semin Avian Exotic Pet Med*. 2005;14:52–60.

5. Pizzi R, Martinez Pereira Y, Feltner Rambaud Y, *et al*. Secundum atrial septal defect in a Komodo dragon (*Varanus komodoensis*). *Vet Rec*. 2009;164:472–473.

6. Jensen B, Wang TW. Hemodynamic consequences of cardiac malformations in two juvenile ball pythons (*Python regius*). *J Zoo Wildl Med*. 2009;40:752–756.

7. Brockman HL, Kennedy JP. Interventricular septal defect in *Alligator mississippiensis* Daudin. *Tex Rep Biol Med*. 1962;20:719–720.

8. Sweet CR, Linnetz E, Golden E, Mayer J. What is your diagnosis? Pericardial effusion, cardiac mass, or cardiomegaly secondary to heart diseases. *J Am Vet Med Assoc*. 2009;234(10):1259–1260.

9. Schuchman SM, Taylor DO. Arteriosclerosis in an iguana (*Iguana iguana*). *J Am Vet Med Assoc*. 1970;157:614–616.

10. Barten S. Cardiomyopathy in a king snake (*Lampropeltis calligaster rhombomaculata*). *Vet Med Small Anim Clin*. 1980;75:125–129.

11. Jacobson ER, Seely JC, Novilla MN, *et al*. Heart failure associated with unusual hepatic inclusions in a Deckert's rat snake. *J Wildl Dis*. 1980;15:75–81.

12. Clippinger TL. Aortic stenosis and atrio/ventricular dilatation in a green iguana (*Iguana iguana*). In: *Proc Am Assoc Zoo Vet*. 1993:342–344.

13. Schilliger L, Paillusseau C, Gandar F, Desprez I, Claude W, Passavin P, Saponaro V, *et al*. Hypertensive heart disease and encephalopathy in a central bearded dragon (*Pogona vitticeps*) with severe atherosclerosis and first-degree atrioventricular block. *J Zoo Wildl Med*. 2019;50(2):482–486.

14. Frye FL. Common pathologic lesions & disease processes. In: *Biomedical and Surgical Aspects of Captive Reptile Husbandry*. Malabar, FL: Krieger Publishing Co.; 1991:560.

15. Jacobson ER. Gentamicin related visceral gout in two boid snakes. *Vet Med Small Anim Clin*. 1976;71:361–363.

16. Montali RJ, Bush M, Smeller JM. The pathology of nephrotoxicity of gentamicin in snakes. *Vet Pathol*. 1979;16:108–111.

17. Jacobson ER, Ginn PE, Troutman JM, *et al*. West Nile virus infection in farmed American alligators (*Alligator mississippiensis*) in Florida. *J Wildl Dis*. 2005;41:96–106.

18. Stenglein MD, Sanders C, Kistler AL, Ruby JG, Franco JY, Reavill DR, Dunker F, DeRisi JL. Identification, characterization, and *in vitro* culture of highly divergent arenaviruses from boa constrictors and annulated tree boas: candidate etiological agents for snake inclusion body disease. *mBio*. 2012;3(4):e00180–12.

19. Bollinger TK, Mao J, Schock D, *et al.* Pathology, isolation, and preliminary molecular characterization of a novel iridovirus from tiger salamanders in Saskatchewan. *J Wildl Dis.* 1999;35(3):413–429.

20. Docherty DE, Meteyer CU, Wang J, *et al.* Diagnostic and molecular evaluation of three iridovirus-associated salamander mortality events. *J Wildl Dis.* 2003;39(3):556–566.

21. Clippinger TL, Bennett RA, Johnson CM, *et al.* Morbidity and mortality associated with a new *Mycoplasma* species from captive American alligators (*Alligator mississippiensis*). *J Zoo Wildl Med.* 2000;31:303–314.

22. Obendorf DL, McManus CJ. *Vibrio damsela* infection in a stranded leatherback turtle (*Dermochelys coriacea*). *J Wildl Dis.* 1987;23:666–668.

23. Jacobson ER, Homer B, Adams W. Endocarditis and congestive heart failure in a Burmese python (*Python molurus bivittatus*). *J Zoo Wildl Med.* 1991;22:245–248.

24. Wernick MB, Novo-Matos J, Ebling A, Kuhn K, Ruetten M, *et al.* Valvulopathy consistent with endocarditis in an Argentine boa (*Boa constrictor occidentalis*). *J Zoo Wildl Med.* 2015;46(1):124–129.

25. Schroff S, Schmidt V, Kiefer I, Krautwald-Junghanns M-E, Pees M. Ultrasonographic diagnosis of an endocarditis valvularis in a Burmese python (*Python molurus bivittatus*) with pneumonia. *J Zoo Wildl Med.* 2010;41(4):721–724.

26. Innis CJ. Myocardial abscess and hemopericardium in a green iguana (*Iguana iguana*). In: *Proc Assoc Rep Amphib Vet.* 2000:185–188.

27. Gray TZ. Cranial body swelling in a boa constrictor, *Boa constrictor constrictor. J Herpetol Med Surg.* 2004;14(4):22–24.

28. Reavill DR, Schmidt RE. Mycobacterial lesions in fish, amphibians, reptiles, rodents, lagomorphs, and ferrets with reference to animal models. *Vet Clin North Am Exot Anim Pract.* 2012;15(1):25–40.

29. Heinz-Taheny KM. Cardiovascular physiology and diseases of amphibians. *Vet Clin North Am Exot Anim Pract.* 2009;12(1):39–50.

30. Schadich E, Cole ALJ. Pathogenicity of *Aeromonas hydrophila*, *Klebsiella pneumoniae*, and *Proteus mirabilis* to brown tree frogs (*Litoria ewingii*). *Comp Med.* 2010;60(2):114–117.

31. Jacobson ER, Gaskin JM, Mansell J. Chlamydial infection in puff adders (*Bitis arietans*). *J Zoo Wildl Med.* 1989;20:364–369.

32. Homer BL, Jacobson ER, Schumacher J, Scherba G. Chlamydiosis in mariculture-reared green sea turtles (*Chelonia mydas*). *Vet Pathol.* 1994;31:1–7.

33. Jacobson E, Origgi F, Heard D, Detrisac C. Immunohistochemical staining of chlamydial antigen in emerald tree boas (*Corallus caninus*). *J Vet Diagn Invest.* 2002;14:487–494.

34. Thongkamkoon P, Tohmee N, Morris EK, Inamnuay L, Lombardini ED. Combined fatal systemic *Chlamydia* sp. and *Aeromonas sobria* infection in juvenile Siamese crocodiles (*Crocodylus siamensis*). *Vet Pathol.* 2018;55(5):736–740.

35. Speare R, Berger L, O'Shea P, Ladds PW, Thomas AD. Pathology of mucormycosis of cane toads in Australia. *J Wildl Dis.* 1997;33(1):105–111.

36. Bube A, Burkhardt E, Weiss R. Spontaneous chromomycosis in the marine toad (*Bufo marinus*). *J Comp Pathol.* 1992;106(1):73–77.

37. Wunschmann A, Armién AG, Childress AL, Wellehan JFX. Intrapericardial *Encephalitozoon pogonae*-associated arteritis with fatal hemopericardium in two juvenile central bearded dragons. *J Vet Diagn Invest.* 2019;31(3):467–470.

38. Gordon N, Kelly WR, Cribb TH. Lesions caused by cardiovascular flukes (Digenea: Spirorchidae) in stranded green turtles (*Chelonia mydas*). *Vet Pathol.* 1998;35:21–30.

39. Cordero-Tapia A, Gardner SC, Arellano-Blanco J, Inohuye-Rivera RB. *Learedius learedi* infection in black turtles (*Chelonia mydas agassizii*), Baja California Sur, Mexico. *J Parasitol.* 2004;90:645–647.

40. Marchiori E, Negrisolo E, Cassini R, Garofalo L, Poppi L, Tessarin C, Marcer F. Cardiovascular flukes (Trematoda: Spirorchiidae) in *Caretta caretta* Linnaeus, 1758 from the Mediterranean Sea. *Parasit Vectors.* 2017;10(1):467.

41. Johnson CA, Griffith JW, Tenorio P, *et al.* Fatal trematodiasis in research turtles. *Lab Anim Sci.* 1998;48:340–343.

42. Glazebrook JS, Campbell RS, Blair D. Studies on cardiovascular fluke (Digenea: Spirorchiidae) infections in sea turtles from the Great Barrier Reef, Queensland, Australia. *J Comp Pathol.* 1989;101:231–250.

43. McAllister CT, Cordes JE, Conn DB, Singleton J, Walker JM. Helminth parasites of unisexual and bisexual whiptail lizards (Teiidae) in North America. V. *Mesocestoides* sp. tetrathyridia (Cestoidea: Cyclophyllidea) from four species of *Cnemidophorus. J Wildl Dis.* 1991;27:494–497.

44. Niebuhr CN, Jarvi SI, Kaluna L, Fischer BLT, Deane AR, Leinbach IL, Siers SR. Occurrence of rat lungworm (*Angiostrongylus cantonensis*) in invasive coqui frogs (*Eleutherodactylus coqui*) and other hosts in Hawaii, USA. *J Wildl Dis.* 2020;56(1):203–207.

45. Shoemaker M, Barrie M, Holman H, Wolk KE, Stromberg PC, Aeffner F. Pathology in practice. Cardiac hemangiosarcoma in a snake. *J Am Vet Med Assoc.* 2016;248(2):153–155.

46. Mauldin CN, Done LB. Oncology. In: Mader DR, ed. *Reptile Medicine and Surgery.* Philadelphia, PA: WB Saunders; 2006:299–322.

47. de Brot S, Sydler T, Nufer L, Ruetten M. Histologic, immunohistochemical, and electron microscopic characterization of a malignant iridophoroma in a dwarf bearded dragon (*Pogona henrylawsoni*). *J Zoo Wildl Med.* 2015;46(3):583–587.

48. Work TM, Balazs GH, Rameyer RA, *et al.* Retrospective pathology survey of green turtles *Chelonia mydas* with fibropapillomatosis in the Hawaiian Islands, 1993–2003. *Dis Aquat Organ.* 2004;62:163–176.

49. Schumacher JR, Bennett A, Fox LE, *et al.* Mast cell tumor in an eastern kingsnake (*Lampropeltis getulus getulus*). *J Vet Diagn Invest.* 1998;10:101–104.

50. Schmidt RE, Reavill DR. Metastatic chondrosarcoma in a corn snake (*Pantherophis guttatus*). *J Herpetol Med Surg.* 2012;22(3–4):67–69.

51. Pereira ME, Viner TC. Oviduct adenocarcinoma in some species of captive snakes. *Vet Pathol.* 2008;45:693–697.

52. Tuttle AD, Harms CA, Grafinger MS, Lewbart GA, Van Wettere AJNJ. Splenic hemangiosarcoma in a corn snake, *Elaphe guttata. J Herpetol Med Surg.* 2006;16(4):140–143.

53. Kubiak M, Denk D, Stidworthy MF. Retrospective review of neoplasms of captive lizards in the United Kingdom. *Vet Rec.* 2020;186(1):28.

54. Stumpel JBG, Del-Pozo J, French A, Eatwell K. Cardiac hemangioma in a corn snake (*Pantherophis guttatus*). *J Zoo Wildl Med.* 2012;43(2):360–366.

55. Gál J, Jakab C, Szabó Z, Pazár P, Psáder R, Roeber F, Hegyi A, Lefler KK, Farkas B, Mándoki M. Haemangioma in the oesophagus of a red-eared slider (*Trachemys scripta elegans*). *Acta Vet Hung.* 2009;57(4):477–484.

Speaker Information
(click the speaker's name to view other papers and abstracts submitted by this speaker)

Ethan Biswell, DVM, DACVP
ZNLabs
Salt Lake City, UT, USA

Drury Reavill, DVM, DABVP (Avian and Reptile/Amphibian Practice), DACVP
ZNLabs
Salt Lake City, UT, USA

URL: https://www.vin.com/doc/?id=9768909

The Reptile Liver: Clinical Investigation and Diagnosis

ExoticsCon Virtual 2020 Proceedings

Stephen J. Divers, BVetMed, DZooMed, DECZM (Herp), DECZM (ZHM), DACZM, FRCVS
Department of Small Animal Medicine and Surgery (Zoological Medicine), College of Veterinary Medicine, University of Georgia, Athens, GA, USA

Abstract

Clinical hepatobiliary disease investigation in reptiles appears deficient given the lack of publications reporting antemortem diagnoses. However, the clinical approach is not dissimilar to that recommended in domesticated species. Detailed history, physical examination findings, clinicopathology, diagnostic imaging, endoscopy, biopsy histopathology, and microbiology are typically able to determine a definitive diagnosis. Definitive diagnosis provides an accurate prognosis and facilitates more appropriate, specific therapy.

Introduction

The reptilian liver is similar in structure and function to that of other vertebrates. It is the largest single visceral organ and may be elongated in snakes and some lizards, or transverse in testudines and other lizards. Liver disease is not uncommon in reptiles, and yet a review of published hepatopathies indicates that the vast majority of cases were diagnosed postmortem. This implies deficiency in the current clinical approach to hepatobiliary disease diagnosis. Therefore, the aim of this presentation is to provide clear guidance on antemortem diagnosis of hepatobiliary diseases in reptiles.[1]

History and Physical Examination

The approach to hepatobiliary disease investigation in reptiles is similar to that in mammals. Recently, the World Small Animal Veterinary Association (WSAVA) published guidelines for the diagnosis of known hepatobiliary diseases in dogs and cats. Although these guidelines relate specifically to dogs and cats, they serve as a useful guide for the appropriate investigation of hepatobiliary disease in reptiles. They also serve as a useful reminder that just because a specific disease process has not been identified in reptiles, it does not mean that it cannot occur. As clinicians involved in the evolution of herpetological medicine, we must continually strive to watch for the first case of a portal systemic shunt or hepatocutaneous syndrome.

Detailed information on performing a physical examination is covered elsewhere. Clinical signs, physical examination findings, and clinicopathologic abnormalities associated with hepatobiliary disease often reflect disruptions in normal structure and function. The liver possesses a large functional reserve, and therefore the appearance of potentially hepatic-associated signs such as icterus and ascites typically represent exhaustion of these reserves that occur late in the course of disease. Therefore, any reptile exhibiting such signs should be investigated as a matter of urgency. Nonspecific signs including intermittent anorexia, vomiting, and lethargy can occur with diseases associated with the liver or a variety of other organ systems. Therefore, diagnostic investigation is also required to differentiate from hepatic and non-hepatic diseases.

Historical events such as the ingestion of known hepatotoxic substances (e.g., aflatoxins, pyrrolizidine alkaloids) or treatment with potentially hepatotoxic drugs (e.g., NSAIDs, glucocorticoids) may suggest the presence of hepatobiliary disease. An obese reptile that becomes anorectic may be predisposed to hepatic lipidosis, while chronic hepatic disease often predisposes to gastrointestinal ulceration, hematemesis, coelomic pain, and melena.

The presence of icterus is the most significant and specific sign associated with hepatobiliary disease. In those squamates that produce bilirubin, a yellow discoloration to the mucous membranes may be obvious, while in other species, a greener pigment (especially in urine/urates) is often appreciated. However, in many cases, mucous membrane pigmentation often complicates evaluation. Commonly reported in mammals, polyuria/polydipsia is seldom appreciated in reptiles; however, coelomic enlargement due to ascites or hepatomegaly, anorexia, poor body condition, neurologic depression, and lethargy might also be indicators of an underlying hepatobiliary disorder.

Clinical Enzymology

Although evaluations of hepatobiliary enzymes (e.g., ALT, AST, ALP, GGT, and SDH) are routine in human and domesticated animal medicine, it would be wrong to assume that they are as reliable in reptiles. All of the commercial enzyme measurements are designed for mammals and are run at 37°C. Very few studies have attempted to objectively measure hepatocellular enzyme activities in healthy reptiles, and even fewer have attempted to demonstrate changes following a hepatic insult. Worse still, very few (if any) of the published studies that this author is aware of have managed to satisfy the requirements of the American Society for Veterinary Clinical Pathologists (ASVCP) for reference range determination. Therefore, clinicians should be cautious of including or excluding hepatobiliary disease on the basis of plasma biochemistry results. Taking the green iguana (*Iguana iguana*) as an example, moderate activities of LDH and AST were found in many tissues, making differentiation between hepatic and non-hepatic (especially muscle) sources impossible, whereas ALT and ALP activities were also generally low but again found in a variety of tissues. Recent research at the University of Georgia demonstrated significant elevations in a number of parameters, compared to controls, following an acute hepatotoxic insult. AST, SDH and LDH levels increased more than 15-fold, and ALT increased fourfold, whereas ALP and GGT did not change significantly.

Non-hepatic tissue trauma (especially involving muscle) can also cause increases in AST and LDH, and these parameters are considered less specific. Enzyme elevations remained high for at least 48 hours after insult but decreased by day 6 in most cases. Therefore, elevation of AST, ALT, LDH and SDH are consistent with the initial phase of acute liver injury, but there is little information on the plasma half-lives of reptile hepatic enzymes. Furthermore, although the degree of enzyme elevations is considered proportional to the degree of hepatobiliary damage, it is not predictive of hepatic function and therefore does not indicate prognosis.

Plasma Proteins

Albumin is exclusively produced by the liver, but because of functional capacity and protein half-life, hypoalbuminemia is typically only seen during chronic hepatic disease. Liver cirrhosis may result in hypoalbuminemia, sodium and water retention, and lead to ascites. Hypoalbuminemia is not specific and can occur secondary to protein-losing nephropathies/enteropathies, exudative cutaneous lesions (e.g., severe burns), vasculitis, acute blood loss, or excessive blood collection. Inadequate nutrition and systemic inflammatory conditions can curtail albumin synthesis. Unfortunately, the measurement of albumin by the bromocresol green dye-binding method has been shown to be inaccurate in turtles when compared to the gold standard of protein electrophoresis. Recent work at the University of Georgia has also demonstrated inaccuracies between albumin measurements by bromocresol green when compared to electrophoresis in the bearded dragon (*Pogona vitticeps*). Based upon current knowledge, the measurement of albumin in chelonians and bearded dragons by bromocresol green is inaccurate. Indeed, until proven otherwise, albumin measurements in all reptiles by anything other than protein electrophoresis should be considered inaccurate and avoided.

The plasma globulin fraction is composed of immunoglobulins and non-immunoglobulins, and hepatic impairment can result in decreases in α- and β-globulins. However, as many immunoglobulins are acute-phase proteins and hepatic production is increased during systemic response to inflammatory disease, hepatitis may actually lead to hypergammaglobulinemia. Some commercial laboratories are able to offer acute-phase protein assays, including C-reactive protein, haptoglobin, and amyloid A, although their accuracy and relevance to reptile diseases have not been proven.

Coagulation abnormalities are common in mammals with severe hepatopathy, but they appear to be rare in reptiles. In reptiles, the extrinsic and common pathways are thought to play major roles in coagulation. The extrinsic pathway can be assessed using prothrombin times (PT); however, the use of commercially available tests using mammalian thromboplastin results in prolonged coagulation times or lack of coagulation. Recent investigations into PT in green iguanas (*Iguana iguana*) and red-eared sliders (*Trachemys scripta elegans*) using reptile-derived thromboplastin have demonstrated clotting times in the range of 22–37 seconds, which are not dissimilar to those reported in endotherms. The value of activated partial thromboplastin time (APTT) in reptiles is likely negligible due to the relative lack of intrinsic pathway clotting factors. Although not completely understood, the impacts of temperature and calcium need to be considered when assessing coagulation in reptiles.

Ammonia

The liver is responsible for detoxifying ammonia of intestinal bacterial origin through the urea cycle. Ammonia is an important, but not exclusive cause of hepatic encephalopathy; however, it is currently the only measurable toxin. Hepatic failure or shunting of portal blood away from the liver can result in hyperammonemia, although in mammals, test sensitivity is greater for shunting disorders. The greatest issue facing testing is sample handling, as blood must be collected into cold heparinized tubes and transported on ice for immediate laboratory evaluation. Currently, there is very little information on the use of ammonia in reptiles.

Uric Acid and Blood Urea Nitrogen

Decreased urea levels have been documented in mammals with severe hepatic fibrosis due to decreased hepatic production. It would also seem likely, therefore, that urea and uric acid levels in ureotelic and uricotelic reptiles would also be decreased due to decreased hepatic production. A recent study investigating hepatotoxin-induced hepatopathy in iguanas reported resting levels of urea and uric acid to be 3.5 mg/dL and 4.1 mg/dL, respectively. Peak levels of 11.5 mg/dL and 4.4 mg/dL were recorded at 96 hours post-insult. These elevations (rather than decreases) emphasize the many confounding variables that can influence urea and uric acid levels, which include hydration status, dietary protein content, recent feeding, gastrointestinal hemorrhage, glomerular filtration rate, drug therapy (e.g., allopurinol), and diuresis.

Bilirubin and Biliverdin

In those snakes that can reduce biliverdin to bilirubin, hepatobiliary disease would be expected to cause an elevation in unconjugated bilirubin. Although elevated bile pigments are more specific for hepatobiliary disease, they are generally less sensitive than hepatocellular enzymes. Hyperbilirubinemia and hyperbiliverdinemia can be prehepatic, hepatic, or posthepatic. Prehepatic causes include severe hemolysis and can be confirmed by demonstrating a decrease in packed cell volume. Hepatic causes can be due to impaired uptake, conjugation, or excretion that cause severe intrahepatic cholestasis. Differentiation between hepatic and posthepatic causes is difficult but important, as posthepatic conditions typically require surgical decompression of the gallbladder, and intrahepatic cases are treated medically. A recent experimental study in iguanas documented resting bilirubin levels of 0.2±0.1 mg/dL, rising to 0.6±0.1 mg/dL 96 hours following the administration of hepatotoxin. However, for most non-serpentine reptiles, measurement of biliverdin is required and, despite the existence of validated assays, few are commercially offered.

Bile Acids

In normal animals during fasting, when the enterohepatic recirculation of bile acids is low, total serum or plasma bile acids are also low. In dogs and cats, bile acids are highly specific for hepatobiliary disease and are an ideal hepatic function test. However, with the exception of portosystemic shunts, the sensitivity of bile acids is insufficient to warrant use as a screening test, and the same may be true for reptiles. In clinically healthy green iguanas and bearded dragons, mean fasting levels of 7.5 (range 2.6–30.3) µmol/L and 4.6 (range 0.8–33.7) µmol/L, respectively, have been documented. Significant postprandial effects have also been demonstrated, and in iguanas, mean elevations to 33.3±22.0 µmol/L have been recorded 3 hours post-feeding. In bearded dragons, mean elevations to 6.9 (range 1.4–37.6) µmol/L and 11.9 (4.3–31.0) µmol/L have been documented at 4 and 24 hours post-feeding. The duration of this postprandial effect has not been determined, but it suggests that reptiles should be fasted for several days prior to sampling, probably longer in large carnivorous reptiles. A number of factors can affect total plasma bile acids, including degree of gallbladder emptying, rate of gastric emptying, intestinal transit rate, efficiency of ileal bile acid reabsorption, frequency of enterohepatic cycling, fat and amino acid content of the test meal, and amount of test meal consumed. Therefore, a consistent approach to performing a bile acid stimulation test is important.

Glucose

Hypoglycemia occurs in mammals with hepatic failure but is rare in cases of chronic disease. Given the already low resting glucose levels of most reptiles and the effects of nutrition and handling stress, the detection of hypoglycemia may be difficult unless accurate baseline values for a specific patient have been repeatedly documented over time. Following hepatotoxin insult to iguanid livers, no significant changes in glucose levels were appreciated over a 6-day period.

Cholesterol, Triglycerides, and Lipids

The liver converts fatty acids into triglycerides, which are either stored or released from the liver as very low-density lipoproteins (VLDL). Cholesterol may be synthesized within the liver or acquired through chylomicron remnants and low-density lipoproteins (LDL). Hepatic cholesterol can be esterified and secreted in lipoproteins or stored in the liver. Hypercholesterolemia can be associated with excessive production or decreased excretion (e.g., extrahepatic biliary obstruction). Normal female reptiles undergoing vitellogenesis often exhibit increases in cholesterol and triglycerides; however, sustained elevations or elevations in males are often seen in cases of hepatic lipidosis. There is a variable postprandial effect on triglyceride levels (less so for cholesterol) in bearded dragons, and fasting prior to sampling is recommended. In addition to cholesterol and triglycerides, some commercial laboratories offer lipoprotein profiles.

Urinalysis

Urinalysis in mammals is useful for the demonstration of ammonia biurate crystalluria and uric acid; however, given the variable expectation for such nitrogenous products in many reptiles, their diagnostic usefulness is probably limited as an adjunct to hepatobiliary disease diagnosis. Bilirubinuria is considered abnormal in cats but not in dogs. Only snakes appear capable of producing bilirubin in significant quantities, but the renal threshold for urinary bilirubin is unknown.

Hematology

Coagulopathies are rare, but bleeding could result in regenerative anemia. Nonregenerative anemia may be present in cases of chronic hepatobiliary disease, and normocytic, normochromic anemia would be expected due to inefficient utilization of systemic iron stores (anemia of chronic disease).

Diagnostic Imaging

Radiography

The size, shape, and position of the liver can be difficult to determine from plain radiographs of the cranial coelom. Unfortunately, due to lack of diffuse fat tissue, serosal detail is typically poor in reptiles, and evaluation of liver margins is often not possible. In addition, although hepatomegaly can cause displacement of the stomach, there is little appreciation of changes in gastric axis in reptiles. Hepatomegaly commonly occurs due to infiltrative disease (e.g., neoplasia, lipidosis, amyloidosis) and inflammatory disease (e.g., bacterial hepatitis). Focal enlargement due to a neoplasm, cyst, granuloma, or abscess may be appreciated by displacement of adjacent tissues and, in squamates, compression on the lungs. Microhepatica, often associated with fibrosis and atrophy (including chronic cachexia), is seen as a decrease in the size of the liver silhouette. Normally the liver is homogeneous in appearance. Mineralization can be diffuse (choledocholithiasis) or as a focal, discrete opacity associated with the right liver lobe in lizards and chelonians, or just caudal to the liver in snakes. Gas opacities may be associated with abscesses or as a grave sign associated with severe necrotizing gastroenteritis.

Computed Tomography (CT)

The size and structure of the liver can be assessed using CT, in much the same way as in mammals. Pre- and post-contrast images in both soft tissue and bone algorithms should be collected. Computed tomography can be used to detect alterations in the x-ray attenuation of hepatic parenchyma. In healthy juvenile sea turtles (*Chelonia mydas*), a mean attenuation value of 60.09±5.3 Hounsfield units (HU) has been reported; in general, values between 50–70 HU are considered normal for chelonians. However, in male captive red-footed tortoises (*Chelonoidis carbonaria*), hepatic values were much lower at 11.2±3.0 HU and were attributed to multiple cases of hepatic lipidosis in the group, although this was not confirmed histologically.

Ultrasonography

Ultrasonography is useful for the evaluation of coelomic viscera, including the liver. However, ultrasonographic anatomy has only been described for a few species, but is generally considered useful for the differentiation between focal and diffuse disease, as well as assessment of displacement/compression, parenchymal changes, gallbladder disease, and vascular abnormalities.

In the green iguana, hepatic parenchyma is of uniform echotexture with medium echogenicity, similar to that of the spleen, testes, and fat bodies. Hepatic lobation is generally indistinct. The gallbladder is easily identified adjacent to and partially embedded in the caudoventral border of the liver, right of midline. Gallbladder contents are normally anechoic, but small amounts of dependent material of mineral echogenicity can sometimes be identified. The subadult iguanid gallbladder is an elongated oval in the sagittal plane, with mean±SD long-axis dimensions of 1.67±0.40 cm (1.00–2.72 cm) and mean short-axis dimensions in the sagittal plane of 0.67±0.16 cm (range, 0.34–1.00 cm). However, significant linear correlations between body weight and gallbladder dimensions appear to be lacking. Preliminary investigations into contrast-enhanced hepatic ultrasonography have indicated that peak enhancement of 19.9±7.5% (11.7–34.6%) is achieved after 134.0±125.1 seconds (59.6–364.5). However, as the distribution of contrast medium in iguanas differs from that of mammals, specific reference ranges of hepatic perfusion for diagnostic evaluation of the reptilian liver are still outstanding.

Ultrasonographic liver anatomy has also been briefly described for the desert tortoise (*Gopherus agassizii*) via left and right mediastinal or cervicobrachial acoustic windows. No descriptions of hepatic parenchymal structure or measurements were given, although visualization of the gallbladder was limited by the amount of gastrointestinal gas present. The liver of the Boa constrictor (*Boa constrictor*) has been described as uniformly heteroechoic, fusiform, and primarily located on the right side of the coelom, with the vena cava ventral and hepatic vein dorsal. The gallbladder, anechoic with hyperechoic wall, is located just caudal to the caudal margin of the liver, in close association with the pancreas, spleen, and fat body.

Magnetic Resonance Imaging (MRI)

The hepatobiliary system can be evaluated using MRI, with most current descriptions based upon chelonians. Image quality is largely dependent upon the size of the animal, strength of the magnet, and scanner software. In general, units capable of generating magnetic fields of 3+ tesla provide better-quality images for most reptiles. The liver is more signal intensive in T1-weighted images compared to those that are T2 weighted. In addition to hepatic size, parenchymal homogeneity can be assessed, and its fat content compared to that of fat bodies. The internal parenchymal structure is interrupted by the vessels, which (along with the biliary system) are better evaluated using T2-weighted images.

Endoscopy

Endoscopic techniques for the evaluation of the chelonian and saurian coelom have been well described. Endoscopic evaluation of the liver provides magnified, color visualization of the hepatic surface and gallbladder, although deeper lesions can often be appreciated, especially if large. In addition, significant lesions can often be identified endoscopically, despite the lack of obvious clinicopathologic, radiographic, ultrasonographic, or cross-sectional imaging abnormalities. In laterally recumbent lizards and chelonians, a lateral paralumbar approach provides visualization of the ventral and dorsal surfaces of the ipsilateral liver lobe. In dorsal recumbency, a paramedian approach provides access to the ventral surface of the entire liver. In green iguanas and red-eared sliders, entry of a 2.7-mm telescope within a 4.8-mm operating sheath and visualization of the liver from both left and right paralumbar approaches were easily performed without complications; however, the gallbladder could only be reliably seen from a right approach. In snakes, a targeted coelioscopic approach is often required, as, in most cases, no single entry can permit examination along the length of the entire liver. However, examination from within the air sac does permit evaluation over a much greater hepatic length.

Fine-Needle Aspirate Cytology

Fine-needle aspiration from the liver is possible under ultrasound or CT guidance. However, due to the loss of tissue architecture, ultrasound-guided cytological aspirates are generally considered inferior to endoscopic or surgical biopsy. A recent study investigated hepatotoxin-induced disease in green iguanas, and invariably ultrasound-guided fine-needle aspirates were of poor diagnostic value. Fine-needle aspirates may be more useful in disease processes that are highly cellular (e.g., neoplasia), for microbiologic evaluation, and where longer periods of general anesthesia carry unacceptable risks.

Biopsy

Clinicopathologic changes may be attributable to hepatobiliary disease; however, they do not reliably differentiate between different disease processes and do not provide an accurate prognosis. To reach a definitive diagnosis, it is essential to demonstrate both (i) an accurate host pathological response and (ii) the etiologic or causative agent. While paired rising titers are definitive, there are few commercial assays available for reptiles, and those that are available typically require samples 6–9 weeks apart. Given the poorer diagnostic value of cytology compared to histopathology, tissue biopsy is generally preferred. Tissue biopsies can be used for histopathologic, microbiologic, parasitologic, and toxicologic evaluations and can typically provide a definitive diagnosis within days of collection. Indications for hepatic biopsy include elevations of hepatic enzymes (especially AST and SDH, bile acids, biliverdin/bilirubin, unexplained coelomic effusion/ascites, and hepatomegaly). The most common complication concerning hepatic biopsy (especially following needle biopsy) is hemorrhage, and therefore consideration should be given to platelet count, PT, PTT, and buccal mucosal bleeding time. Biopsy samples may be collected using cutting biopsy needles (e.g., Tru-Cut), endoscopically, or by standard wedge or excisional surgical techniques.

Handling of biopsies should, whenever possible, be undertaken with atraumatic instruments to minimize damage. Small endoscopic and needle biopsies should be gently shaken free into a sterile saline and not removed using cotton-tipped applicators or, worse still, a hypodermic needle. Small samples should be placed into labeled histology filters before being submitted in 10% neutral buffered formalin for histopathology. For transmission electron microscopy (TEM), glutaraldehyde is often used, while ethanol or methanol may be preferred for urate tophi or parasite evaluations. Tissue samples for microbiology should be sent in appropriate transport media depending upon whether fungal, bacterial, or viral investigations are most important. Biopsies can be frozen at -20°C while histopathology results are pending. Then, if histopathologically indicated, such frozen tissue can be submitted for specific microbiologic (often virologic) investigations.

Ultrasound-guided, percutaneous hepatic biopsies have been reported from 1–25-kg boid snakes using an 18-gauge x 20-cm biopsy needle. All collected biopsies were considered diagnostic, and all snakes survived; however, multiple biopsy attempts were required for some snakes, and iatrogenic gastric perforation was documented in at least one of 15 animals (7%). Ultrasound-guided liver biopsy in green iguanas has also been described, but one of eight animals (12.5%) suffered from fatal hemorrhage; the technique is probably unsafe, especially for smaller lizards.

Although general anesthesia is seldom employed for such biopsies in humans and dogs, it is still recommended for reptiles to ensure adequate restraint. Automated needles are preferred and operated by the ultrasonographer to ensure that vessels and other organs are not in the path of the needle. Holding an anesthetized reptile at maximum inspiration tends to improve localization and visualization of the liver. The biopsy site should be evaluated for post-sample hemorrhage. Comparison between needle biopsies and wedge biopsy (gold standard) in dogs and cats has indicated an overall agreement on only 48%, and similar results would likely be expected for reptiles.

Endoscopic liver biopsies have been experimentally evaluated in green iguanas and freshwater chelonians. Biopsies were collected using 1.7-mm biopsy forceps, and each contained a mean of 1.9 and 3.1 portal tracts in iguanas and sliders, respectively. Although peripheral crush artifact was common, it was minimal, and all samples were considered diagnostic. No adverse effects of endoscopic hepatic biopsy were reported. Multiple samples can be readily collected and should be routinely submitted for histopathology, bacterial (aerobic/anerobic) and fungal cultures. Additional parasitic or toxicologic analysis can also be considered on an individual case basis. This technique enables gross evaluation of most (if not all) of the liver, extrahepatic biliary system, and surrounding structures. The ability to collect multiple samples from various hepatic locations decreases the risks of sampling artifact in cases of regional diversity within the liver. The standard oval/round cup biopsy forceps also produces less hemorrhage than biopsy needles; but if hemorrhage is encountered, direct pressure can be applied by the closed biopsy forceps or a blunt probe, or radiosurgical coagulation can be used. Direct visualization of the hepatic parenchyma enables the clinician to correlate clinical data with liver appearance and histopathology to adjudicate the most accurate diagnosis.

Coeliotomy provides access to the liver for harvesting surgical biopsies. The advantages of open surgical access are the ability to collect larger liver samples, as well as an improved opportunity to more completely evaluate and sample other visceral structures (e.g., intestines). Nevertheless, increased invasion and duration of anesthesia must be considered. Surgical biopsies can be collected using a standard wedge technique. Hemorrhage can be controlled using hemostatic material (e.g., Gelfoam) and digital pressure. Advantages and disadvantages are similar to those of endoscopy.

References

1. Divers SJ. Hepatology. In: Divers SJ, Stahl SJ, eds. *Mader's Reptile and Amphibian Medicine and Surgery*. 3rd ed. St. Louis, MO: Elsevier; 2019:649–668.

An extensive list of references can be found in the above review chapter.

SPEAKER INFORMATION
(click the speaker's name to view other papers and abstracts submitted by this speaker)

Stephen J. Divers, BVetMed, DZooMed, DECZM (Herp), DECZM (ZHM), DACZM, FRCVS
Department of Small Animal Medicine and Surgery (Zoological Medicine)
College of Veterinary Medicine
University of Georgia
Athens, GA, USA

URL: https://www.vin.com/doc/?id=9768907

Reptilian Cardiopulmonary Physiology and Inhalant Anesthesia: What Do We Know So Far?

ExoticsCon Virtual 2020 Proceedings

Bonnie Gatson, DVM, DACVAA

Department of Anesthesia and Pain Management, College of Veterinary Medicine, University of Florida, Gainesville, FL, USA

ABSTRACT

The cardiovascular and pulmonary systems of non-avian reptiles are diverse and complex, making anesthetic management of these species challenging. Inhalant anesthetics are commonly chosen to induce and maintain general anesthesia in many reptilian species for prolonged diagnostic and surgical procedures. Although these anesthetic agents are routinely celebrated for their ease of use and rapid ability to titrate the depth of anesthesia in mammalian species, it has been well documented that induction and recovery with these agents in reptiles are both prolonged and unpredictable. The cause of this is likely multifaceted and affected by factors such as optimal body temperature of the species, degree of cardiopulmonary circulation, and changes in ventilation that occur as a result of general anesthesia. Understanding the impact these various factors have on the speed of induction and recovery using inhalant agents in reptiles will allow practitioners and anesthetists to appropriately alter the course of inhalant anesthesia across different reptilian species, with a goal of improving overall anesthetic management.

General anesthesia is a necessity in veterinary practice to provide unconsciousness, analgesia, and muscle relaxation for surgical and diagnostic procedures. It is commonly utilized for these purposes in reptilian medical and surgical practices, with the added benefit of facilitating safe handling of dangerous species. Volatile inhalant agents are generally favored for induction and maintenance of general anesthesia, as they allow careful titration of anesthetic depth and are reversible, allowing a smooth transition to consciousness once administration is discontinued. Although these advantageous properties are often described for mammals and birds, these benefits do not necessarily cross over to other species, particularly reptiles.

Use of inhalant anesthetics, specifically isoflurane and sevoflurane, has been described in chelonians, snakes, and varanid and non-varanid lizards. For these species, inhalant agents have been associated with a slow induction of anesthesia, an inadequate anesthetic plane, as well as delayed or unpredictable recoveries.[1] Likewise, the minimum alveolar concentration (MAC) of isoflurane reported in reptiles is extremely variable compared to what is reported in other species, where MAC is generally well conserved across mammalian and avian species.[1] There are a variety of factors contributing to these species differences, including a unique cardiovascular and thermoregulatory physiology that alters the traditional understanding of inhalant pharmacokinetics in reptiles.

Modern inhalants including isoflurane, sevoflurane, and desflurane are a distinctive class of drugs, as they are vaporized and administered through the respiratory system to be absorbed into the bloodstream and carried to the central nervous system, where they exert their effects. As inhalants are vapors, these agents behave in the body like gases, and their pharmacokinetic properties are heavily influenced by this fact. The speed of anesthetic induction using inhalant agents is dependent on a balance between the rate of delivery of anesthetic into the lungs and the uptake of anesthetic away from the lungs into pulmonary circulation.[2] Delivery of volatile anesthetics into the lungs is dependent upon creating an appropriate driving pressure from the anesthetic machine into the patient's respiratory system as well as the rate of pulmonary ventilation.[2] Uptake of anesthetic into pulmonary circulation is dependent on several major factors: the creation of an adequate pressure differential between the lungs and pulmonary circulation to drive inhalants into the bloodstream, the ability of inhalants to dissolve into the bloodstream (known an inhalant solubility), and the animal's cardiac output.[2] Inhalant solubility is heavily influenced by body temperature, as the colder the patient becomes, the more the gaseous agent will dissolve into the bloodstream. Lower blood solubility is associated with a faster anesthetic induction, as less inhalant needs to be dissolved into the bloodstream to be carried to the brain and spinal cord to exert its effects.[2] Recovery from inhalant agents is also dependent on these factors, as they undergo minimal hepatic metabolism and elimination is dependent on pulmonary ventilation.[2] However, the duration of general anesthesia plays a significant role with certain inhalant agents, such as isoflurane.[2] Factors that directly affect anesthetic induction and recovery using inhalant agents can be altered significantly by the unique physiologic properties of reptiles.

All reptiles can shunt blood away from pulmonary circulation. Non-crocodilian reptiles possess a single ventricle where mixing of oxygenated and deoxygenated blood occurs during diastole. The degree of intracardiac right-to-left shunting is influenced by a variety of overlapping factors, including pressure differentials between the systemic and pulmonary circulation, ventilation patterns, and the amount of blood remaining in the cavum dorsale at the end of systole.[1,3-6] Crocodilian shunting patterns are similarly complex even though they possess a completely separated ventricle similar to mammals and birds. Extracardiac right-to-left shunting is facilitated through several mechanisms. In crocodilians, the right aorta exits from the left ventricle while the left aorta comes from the right ventricle along with the main pulmonary artery. Two connections exist between the right and left aortas that facilitate extracardiac shunting: one at the base of the heart, known as the foramen of Panizza, and another connection known as the abdominal anastomosis.[1,7,8] An additional shunting mechanism exists within the subpulmonary conus located within the pulmonary outflow tract in the right ventricle that contains a cog-teeth valve.[8] The diameters of both the subpulmonary conus and foramen of Panizza can change in size depending on a variety of neurohormonal substances and adrenergic influences.[7,8] As the diameter of the subpulmonary conus narrows, a greater fraction of blood is directed out of the left aorta, thereby bypassing pulmonary circulation.[8]

Although a variety of shunting patterns have been documented in both non-crocodilians and crocodilians, in theory, bypassing pulmonary circulation can significantly slow the rate of both induction and recovery using inhalant agents. Although a variety of mechanisms are involved in controlling cardiovascular shunting patterns in reptiles, alteration in autonomic drive to the pulmonary or systemic circulatory systems is one proven mechanism of physiologic shunting.[3-6] In the red-eared slider turtle (*Trachemys scripta elegans*), increasing parasympathetic tone results in greater pulmonary vascular resistance, subsequently decreasing pulmonary blood flow.[3] The opposite occurs in the face of increasing sympathetic tone, where intracardiac shunting patterns favor pulmonary outflow and pulmonary vascular resistance reduces, resulting in greater pulmonary blood flow.[3] Pharmacologic manipulation of the autonomic nervous system shows potential to hasten anesthetic recovery and provide a stable plane of anesthesia in some chelonian species. In the common snapping turtle (*Chelydra serpentina*), intramuscular administration of epinephrine at 0.1 mg/kg during recovery from isoflurane resulted in faster return to spontaneous ventilation, spontaneous movement, and extubation times while in loggerhead sea turtles (*Caretta caretta*), intramuscular epinephrine results in a significantly faster return of spontaneous movement.[4,5] Pulmonary blood flow was significantly increased using intravenous atropine at 1 mg/kg in red-footed tortoises (*Chelonoidis carbonaria*), resulting in a decrease in isoflurane concentration required to maintain a light plane of anesthesia in this species.[6]

Although intracardiac right-to-left shunting does not occur in crocodilians, the diameter of extracardiac structures involved in regulating vascular shunting, such as the foramen of Panizza and the subpulmonary conus, are under autonomic control.[7,8] Therefore, similar to chelonians, the administration of intramuscular epinephrine at 0.1 mg/kg during recovery from isoflurane anesthesia in American alligators (*Alligator mississippiensis*) significantly hastens the time to spontaneous ventilation and the time to extubation.[9] Echocardiographic evaluation of American alligators after epinephrine administration shows a complete closure of the foramen of Panizza and an overall increase in pulmonary outflow.[10] Although no long-term adverse effects were noted in these studies following a single dose of epinephrine or atropine, other studies in alligators have linked epinephrine administration to a significant increase in intracranial pressure.[11] Therefore, the use of epinephrine to hasten anesthetic recovery may be detrimental for animals with intracranial disease. The degree of vagal tone has also been shown to affect cardiac shunting patterns in the South American rattlesnake (*Crotalus durissus*).[12] However, pharmacologic manipulation of the autonomic nervous system to alter inhalant anesthetic recovery has not been documented in snakes. It should be noted that pythons, in particular, have a low capacity for significant right-to-left intracardiac shunting due to their unique cardiac anatomy compared to other squamates.

The rate of ventilation and the efficiency of gas exchange play a critical role in inhalant induction and elimination. Pulmonary physiology and central control of ventilation vary significantly between reptilian species. Generally speaking, reptiles have a larger tidal volume but lower respiratory surface area compared to similarly sized mammals.[1] However, varanids and other reptiles with higher metabolic demands tend to have a higher respiratory surface area than snakes or chelonians.[1] Various breathing patterns exist across different species. While many reptiles exhibit episodic breathing, aquatic species can hold their breath for prolonged periods of time.[1] The amount of pulmonary blood flow can be intermittent depending on these breathing patterns.[1,5] Pulmonary perfusion and heart rate have been shown in chelonians to increase significantly during spontaneous ventilation, and chelonians, crocodilians, and lizards possess the ability to decrease right-sided perfusion during periods of pulmonary hypoxia.[13] While the physiologic goals are to maintain appropriate ventilation and perfusion matching during various degrees of pulmonary ventilation, the introduction of general anesthesia can cause significant changes to these breathing patterns. For example, both inspiration and expiration are active processes in reptiles, and in some species ventilation is invariably linked to locomotion.[1] The elimination of these active processes as a result of muscle relaxation with inhalant anesthesia can contribute to ventilatory depression, which can slow the rate of anesthetic induction or recovery. In snakes and lizards, increasing carbon dioxide tensions (which can occur as a result of ventilatory depression with general anesthesia) can decrease minute ventilation further.[1]

In order to stabilize gas exchange, support ventilation, and prevent hypoxemia, mechanical ventilation is often performed during inhalant anesthesia. Mechanical ventilation is oftentimes used with inhalant agents to stabilize the anesthetic plane, as it allows for uninterrupted delivery of inhalant agents to the pulmonary system. In mammals, the positive pressure conferred into the thoracic cavity during the inspiratory cycle is linked to an initial increase in pulmonary flow as blood is compressed into the pulmonary artery, followed by a subsequent drop to pulmonary outflow from an increase in pulmonary vascular resistance due to transluminal compression of these low-resistance vessels during the subsequent cardiac cycle. Therefore, although mechanical ventilation ensures continuous delivery of inhalants, variations in pulmonary outflow throughout the ventilatory cycle can affect uptake into pulmonary circulation. Although mechanical ventilation is recommended during inhalant anesthesia of all reptiles, its effect on inhalant kinetics among various reptilian species is not well known. In the pond slider (*Trachemys scripta*), mechanical ventilation had no impact on pulmonary or systemic blood flow.[14] In ball pythons (*Python regius*), mechanically ventilating to high minute volumes (125–250 ml/kg/min) during 60 minutes of isoflurane anesthesia resulted in significantly faster recovery times—possibly due to faster washout of the anesthetic agent from the snakes and ventilation system.[15]

Inhalants are frequently delivered to patients combined with oxygen as a carrier gas. Breathing high concentrations of oxygen has been associated with respiratory depression in snakes and chelonians and may contribute to slowing the rate of anesthetic recovery in these species. For example, in green iguanas (*Iguana iguana*), ventilating with room air (21% oxygen) rather than 100% oxygen in the post-anesthetic period was associated with a faster anesthetic recovery.[1] However, the use of high concentrations of oxygen to deliver inhalants had no significant effect on the duration of anesthetic induction or recovery in various species of lizards, including the bearded dragon (*Pogona vitticeps*) and the Dumeril's monitor (*Varanus dumerilii*).[1,16]

The efficiency of gas diffusion from the functional lung unit into pulmonary circulation may play a small role in the speed of anesthetic induction and recovery, as modern inhalants readily cross this barrier. While mammals possess a bronchoalveolar tidal volume system, it has been recently shown that crocodilians, as well as the Savannah monitor lizard (*Varanus exanthematicus*), evolved a unidirectional pulmonary airflow system similar to the avian respiratory system.[17] Conventional wisdom suggests that this airflow system creates a cross-current for gas exchange that would result in less work of breathing as well as higher gas extraction ratios. The efficiency of this type of respiratory system and its significance when using inhalant anesthetic agents can be demonstrated in birds, as many flying avian species undergo relatively rapid induction and recovery when modern inhalant agents are used alone for general anesthesia. However, for reptilian species with relatively low metabolic rates and oxygen consumption rates, the evolutionary significance of a unidirectional respiratory system is speculative at best. Recent studies in crocodilians have demonstrated that the unidirectional airflow system allows for continued flow through gas exchange parenchyma facilitated by the beating heart even during periods of prolonged apnea.[17] At this point, the impact of this unidirectional airflow system on inhalant kinetics in reptiles is unknown.

Although modern inhalants undergo minimal metabolism prior to elimination, the underlying metabolic rate has a significant impact on anesthesia in reptiles. The metabolic rate of reptiles varies between species and can be altered with changes in body temperature and physical activity, during states of hibernation, and between meals.[1] As reptiles are ectotherms, cardiac shunting plays a significant role in body temperature regulation. Right-to-left shunting facilitates warming, as it increases the fraction of blood flow directed to the periphery for heating before returning to the core.[1] Increasing body temperature has also been documented to increase heart rate and cardiac output, which can facilitate recovery from inhalant anesthetics.[1] Although body temperature has been shown to affect the onset and duration of injectable anesthetic agents in both chelonians and lizards, the impact of body temperature on the onset, duration, and recovery from inhalant anesthetic agents has not been evaluated in reptilian species.[18] As previously mentioned, inhalant solubility, or the ability of inhalants to saturate into the blood from a gaseous state, is altered by body temperature. Inhalant agents with a low solubility in blood, such as sevoflurane, are associated with a faster rate of induction and recovery than agents like isoflurane with a higher blood solubility.[2] In the green iguana, differences in the blood solubility between isoflurane and sevoflurane had no impact on the speed of induction or recovery with these agents.[19] However, the influence of altering body temperatures on the speed of induction and recovery with these two different agents was not evaluated in this species.

Inhalant anesthetic agents are vital to the practice of anesthesia in reptiles. However, a thorough understanding of their pharmacokinetic application to an individual reptilian species is important to utilize these agents appropriately. Although recent studies have shown promise in the use of sympathomimetic agents to speed anesthetic recovery from inhalant agents, more studies are still needed to minimize the unpredictability of induction and recovery when utilizing inhalant anesthetics in various reptilian species.

References

1. Mosley CAE. Anesthesia and analgesia in reptiles. *Semin Avian Exot Pet.* 2005;14(4):243–262.

2. Steffey EP, Mama KR, Brosnan RJ. Inhalation anesthetics. In: Grimm KA, Lamont LA, Tranquilli WJ, Greene SA, Robertson SA, eds. *Veterinary Anesthesia and Analgesia: The Fifth Edition of Lumb and Jones.* Ames, IA: John Wiley & Sons, Inc.; 2015:297–331.

3. Hicks JW, Comeau SG. Vagal regulation of intracardiac shunting in the turtle *Pseudemys scripta. J Exp Biol.* 1994;286(1):109–126.

4. Balko JA, Gatson BJ, Cohen EB, *et al.* Inhalant anesthetic recovery following intramuscular epinephrine in the loggerhead sea turtle (*Caretta caretta*). *J Zoo Wildl Med.* 2018;49(3):680–688.

5. Goe A, Shmalberg J, Gatson B, *et al.* Epinephrine or GV-26 stimulation reduces inhalant anesthetic recovery time in common snapping turtles (*Chelydra serpentine*). *J Zoo Wildl Med.* 2016;47(2):501–507.

6. Greunz EM, Williams C, Ringgaard S, *et al.* Elimination of intracardiac shunting provides stable gas anesthesia in tortoises. *Sci Rep.* 2018;8(1):17124.

7. Axelsson M, Franklin CE. The calibre of the foramen of Panizza in *Crocodylus porosus* in variable and under adrenergic control. *J Comp Physiol B.* 2001;171:341–346.

8. Axelsson M, Franklin CE, Fritsche R, *et al.* The sub-pulmonary conus and the arterial anastomosis as important sites of cardiovascular regulation in the crocodile *Crocodylus porosus. J Exp Biol.* 1997;200:807–814.

9. Gatson BJ, Goe A, Granone TD, *et al.* Intramuscular epinephrine results in reduced anesthetic recovery time in American alligators (*Alligator mississippiensis*) undergoing isoflurane anesthesia. *J Zoo Wildl Med.* 2017;48(1):55–61.

10. Furst NR, Gatson BJ, Swift S, *et al.* The influence of intramuscular epinephrine on cardiovascular flow patterns in American alligators (*Alligator mississippiensis*) during recovery from isoflurane anesthesia. *Vet Anaesth Analg.* 2018;45(6):885E4–E5.

11. Kondrashova T, Blanchard J, Knoche L, *et al.* Intracranial pressure in the American alligator (*Alligator mississippiensis*): reptilian meninges and orthostatic gradients. *J Comp Physiol A Neuroethol Sens Neural Behav Physiol.* 2020;206(1):45–54.

12. Filogonio R, Wang T, Taylor EW, *et al.* Vagal tone regulates cardiac shunts during activity and at low temperatures in the South American rattlesnake, *Crotalus durissus. J Comp Physiol B.* 2016;186(8):1059–1066.

13. Skovgaard N, Abe AS, Andrade DV, *et al.* Hypoxic pulmonary vasoconstriction in reptiles: a comparative study of four species with different lung structures and pulmonary blood

pressures. *Am J Physiol Regul Integr Comp Physiol*. 2005;289(5):R1280–R1288.

14. Herman J, Wang T, Smits AW, *et al*. The effects of artificial lung inflation on pulmonary blood flow and heart rate in the turtle *Trachemys scripta*. *J Exp Biol*. 1997;200:2539–2545.

15. Jakobsen SL, Williams CJA, Wang T, *et al*. The influence of mechanical ventilation on physiological parameters in ball pythons (*Python regius*). *Comp Biochem Physiol A Mol Integr Physiol*. 2017;207:30–35.

16. Odette O, Churgin SM, Sladky KK, *et al*. Anesthetic induction and recovery parameters in bearded dragons (*Pogona vitticeps*): comparison of isoflurane delivered in 100% oxygen versus 21% oxygen. *J Zoo Wildl Med*. 2015;46(3):534–539.

17. Farmer CG. The evolution of unidirectional pulmonary airflow. *Physiology (Bethesda)*. 2015;30(4):260–272.

18. Kischinovsky M, Duse A, Wang T, *et al*. Intramuscular administration of alfaxalone in red-eared sliders (*Trachemys scripta elegans*)—effects of dose and body temperature. *Vet Anaesth Analg*. 2013;40(1):13–20.

19. Brosnan RJ, Pypendop BH, Barter LS, *et al*. Pharmacokinetics of inhaled anesthetics in green iguanas (*Iguana iguana*). *Am J Vet Res*. 2006;67(10):1670–1674.

SPEAKER INFORMATION
(click the speaker's name to view other papers and abstracts submitted by this speaker)

Bonnie Gatson, DVM, DACVAA
Department of Anesthesia and Pain Management
College of Veterinary Medicine
University of Florida
Gainesville, FL, USA

URL: https://www.vin.com/doc/?id=9768906

Reptile Emergencies

ExoticsCon Virtual 2020 Proceedings

Heather A. Bjornebo, DVM, DABVP (Reptile and Amphibian Practice), CertAqV
Arizona Exotic Animal Hospital, Mesa, AZ, USA

There are few true reptile emergencies which include trauma, toxicity, or envenomation. Reptiles often present critical on emergency with previously unaddressed, chronic illness. Seizures, hemorrhage, organ prolapse, severe trauma, dyspnea, and loss of responsiveness are all clinical presentations that warrant emergency evaluation in reptile patients. Triage should occur on presentation, and a thorough history should be obtained along with a thorough physical examination, and various diagnostic modalities should be utilized to further assess the critical reptile patient. Providing supportive care along with appropriate treatment is paramount to patient recovery.[1,2]

Introduction

It is often said there are few true reptile emergencies. True, acute emergencies often occur in cases of trauma, toxicity, or envenomation. However, it is common for reptiles to present on emergency after a protracted course of illness that either has gone unaddressed or unnoticed by the owner. Patient status may be quite critical in these cases. Seizures, hemorrhage, organ prolapse, severe trauma, dyspnea, and loss of responsiveness are all clinical presentations that warrant emergency evaluation in reptile patients.[1,2]

Assessment, Triage, and Diagnostics

A thorough history should be obtained for any reptile presenting for physical examination and should include a review of where the animal was obtained, previous medical treatment, and current husbandry. Husbandry should be compared to that species' natural environment, diet, and behaviors found in the wild, as deficits in meeting the patient's husbandry needs often contribute to the onset of disease. Correction of any deficiencies will be necessary for resolution of the disease.[2] As with mammalian patients, information such as diet, abnormal behaviors, or weight changes should be recorded.[1,2]

It is also important to determine the temperature ranges provided in the reptile's environmental home enclosure. Reptiles are poikilothermic and should be kept within their preferred optimal temperature zone (POTZ) while hospitalized, to allow for best possible metabolic and immune function. Patients whose body temperature is low on arrival to the hospital should be warmed to within their POTZ slowly over 4–5 hours. Acceptable methods include warming blankets, heated air blankets, heat lamps, or incubators. Many physiologic parameters are best evaluated after the patient has reached optimal temperature.[1,2]

Mucous membrane color and capillary refill time are not accurate assessments of cardiovascular function in reptile patients. Indirect blood pressure measures are often inconsistent and inaccurate, and direct measurement is often too invasive and not practical for clinical use. The best means to assess the cardiovascular system is by determining heart rate. However, cardiac size and anatomy in reptiles often inhibit auscultation with a standard stethoscope. Ultrasonic Doppler flow detectors directed over the heart or major arteries are necessary to evaluate the heart in most reptile patients. Larger lizards and crocodilians may require ultrasonography to assess cardiac function. It should be noted that presence of a heartbeat does not necessarily indicate a reptile patient is alive—their hearts may beat for hours after brain function ceases. It is important to record baseline heart rates once the patient is within their POTZ, because heart rate is highly dependent on body temperature in reptiles.[1,2]

Normal respirations for reptiles can be slow and intermittent. Observe the patient prior to handling, as stress can greatly influence respiratory rate and character. Dyspnea in reptiles can be recognized often by exaggerated respiratory movements. A dyspneic patient may open-mouth gasp with each breath or display extension of the head, neck, and forelimbs. In the event of respiratory arrest, immediately evaluate heart function with a Doppler flow meter and proceed with cardiopulmonary cerebral resuscitation (CPCR) if necessary.[1,2]

It is important to assess mentation and neurologic status; sick reptiles often present with lethargic to dull mentation. Righting reflexes can be determined by placing lizards and snakes into dorsal recumbency. In patients willing to ambulate, gait should be observed. Evaluating reptile patients for pain can be challenging for the practitioner, as behavioral signs of pain tend to be far subtler than in mammals or birds. As a general rule, any condition that would be considered painful in higher vertebrates should be considered painful in reptiles, and analgesia should be provided appropriately.[1]

Hydration status can be subjectively evaluated on examination. Dehydration in reptiles can manifest as a decrease in skin elasticity, decreased skin and eyelid turgor, eyes appearing sunken to the orbits, tacky mucous membranes, doughy coelom on palpation, and thick, stringy saliva. Blood collection provides a more objective evaluation of hydration status. Dehydration may manifest with increases in packed cell volume, total protein, uric acid, blood urea nitrogen (chelonians and crocodilians), sodium, and chloride. Urine specific gravity is of little value in the dehydrated reptile patient, as these species lack the ability to concentrate urine above plasma osmolality.[1]

As with mammalian patients, the importance of blood collection and evaluation to the reptile emergency patient goes beyond evaluation of hydration status. Normal blood pH in reptiles ranges between 7.5 and 7.7, and total CO_2 ranges between 20 and 30 mmol/L. These can be used to assess for metabolic acidosis, which may indicate renal impairment, hypovolemia, or diarrhea. However, as with many physiologic processes in these species, the acid/base system in reptiles is highly dependent on environmental temperature. Temperatures inversely impact blood pH—lower temperatures cause a transient alkalosis—so patient warming is again vital.[1]

Additional blood chemistry parameters that may assist in assessment of the patient include enzyme assays, glucose, calcium, and phosphorus levels. Unlike many mammal patients, enzyme activities have not been found to be organ specific. Many analytes are found in various tissues throughout the body. Interpretation of abnormal parameters requires careful deduction by the practitioner.[1,3]

Reptiles generally have lower blood glucose levels than those of mammals and birds. Hypoglycemia associated with specific disease processes is rarely documented. Hyperglycemia can be associated with stress, postprandial status after starvation, emergence from hibernation, breeding season (temperate reptiles), cryoprotection, forced submergence in turtles, vomiting (hypochloremic metabolic acidosis), epinephrine release, pancreatic disease, liver disease (hepatic lipidosis), renal disease, neoplasia (liver, pancreas, GI, and undifferentiated tissues), and granulomatous disease. Severe hyperglycemia (1000+ mg/dl) is associated with gastric neuroendocrine carcinomas in bearded dragons (*Pogona vitticeps*).[1,3]

Calcium and phosphorus blood levels in reptiles can be indicative of either nutritional problems or significant renal impairment. Mild hypercalcemia is also seen in reproductively active females, with severe hypercalcemia commonly observed in females with reproductive pathology.[1,3]

Hematology can be very beneficial to evaluating the emergency reptile patient. For the most part, indications for and the basic principles of hematology in reptiles are similar to those for mammalian patients. However, the presence of nucleated red blood cells prevents use of automated analyzers and requires manual complete blood counts. Practitioners wishing to perform this diagnostic in-house should familiarize themselves with the reptile blood cell morphology. Diagnosis of anemia should be made with caution, as lymph vessels run in tandem with veins, and contamination during sample collection causes lymphodilution. True anemias are often nonregenerative secondary to chronic illness unless resulting from acute blood loss. Polycythemia is rare and most often indicates dehydration. Leukocytosis in reptiles indicates an inflammatory process. However, absence of a reactive leukogram does not indicate lack of infection. It is common for leukocyte counts to return to normal in reptiles suffering from chronic infections. In many cases, clinical interpretation will rely on evaluation of cell ratios within the differential itself instead of overall leukocyte numbers alone.[1]

Diagnostic imaging can be a valuable tool in the evaluation of the emergency reptile patient. Just as with mammals, orthogonal radiographic views are recommended. Lateral radiographs should be taken with a horizontal beam if possible. Ultrasound can be especially useful in evaluation of effusions.[1]

Therapeutics

With all veterinary patients, good supportive care is key to any therapeutic plan. For poikilotherms, environmental optimization is paramount. As discussed above, returning the patient to the appropriate POTZ is vital for their recovery. Although reptiles do not have an endothermic fever response like mammals, they are known to increase their body temperature behaviorally during infections by basking at the higher end of their POTZ.[1]

Fluid therapy is indicated for patients that are dehydrated or debilitated. It is important to consider that reptiles carry a higher portion of their body fluid intracellularly versus extracellularly, with plasma only accounting for 3.3–7% of total body weight.[1] Reptile osmolality is similar to mammals, and similar fluid types can be used.[1,2] Maintenance fluid requirements in reptiles range from 10–30 ml/kg/d. Fluids should be warmed prior to administration.[2]

For reptiles that are only mildly dehydrated (<5%), fluids are typically provided through the enteral route. Enteral fluid administration may be accomplished passively by soaking the patient in shallow, warm water for 15–30 minutes or actively by gavaging fluid directly into the stomach. No more than 2–3% of body weight should be given in a single enteral dose, as larger volumes may result in regurgitation.[1]

Parenteral fluids are recommended for patients with more severe dehydration (>5%). Options include subcutaneous, epicoelomic, intracoelomic, intravenous, and intraosseous routes. It is recommended to limit the use of the subcutaneous route to patients <6% dehydrated, as the subcutaneous space is relatively small and poorly vascularized in reptiles. The epicoelomic route is ideal for fluid administration of all but the most debilitated chelonian patients; 1–2% of body weight can easily be administered. The epicoelomic route is ideal for hospitalized chelonians, as there is some evidence that the epicoelomic space may be connected to the pericardium and other fluid storage sites, facilitating rapid absorption. That said, the intracoelomic space is lined with a highly absorptive membrane and serosal surfaces and should be considered as an option in saurians. However, fluid administration volume is limited, as excess intracoelomic fluid can result in lung compression. As with higher vertebrates, intravenous fluid administration is the most effective method of fluid administration, but venous access can be challenging due to patient size and thick skin requiring surgical cutdown for access. In severely debilitated patients—especially when intravenous access is not possible—the intraosseous route should be considered for fluid administration. Use shorter needles and confirm placement radiographically. Large fluid boluses can be difficult to administer through intraosseous catheters due to high back pressure.[1]

Increased temperatures also increase the caloric needs of the patient, so it is vital to provide adequate nutritional support. Patients that remain appetent should be offered a diet appropriate for the species. Many debilitated reptiles suffer from some degree of anorexia, maldigestion, and malabsorption—necessitating assist-feeding by syringe. For patients that cannot or will not swallow, a gavage needle or red rubber catheter can be used to directly introduce food into the stomach. Care should be taken to prevent damage to the esophagus or stomach. Some patients, such as medium to large chelonians, will not tolerate restraint for assist-feeding; an indwelling feeding tube should be considered. Local or general anesthesia is required for placement. Placing a pharyngostomy or esophagostomy tube in reptiles is relatively straightforward and extremely similar to placement in mammalian patients. As with mammals, take care to ensure the tube is appropriately placed within the esophagus. Either directly visualize the tube or pull it out through the mouth before redirecting it down the esophagus. The tube should extend into the stomach, and its position should be confirmed radiographically. It is important that the tube is not placed too deeply; excessive depth can cause pressure necrosis of the stomach wall. The tube should be secured to the skin, as with mammal patients, and then further secured in a way that the animal cannot easily dislodge it. It will need to remain in place until the patient's appetite and weight have returned to normal. Once removed, the stoma can be allowed to heal via second intention.[1,4]

High-quality commercial assist-feeding diets are available and have been specifically formulated for herbivores, carnivores, omnivores, and piscivores. Selection should depend on the species' dietary requirements. These diets are stored as a powder and reconstituted with water prior to administration. These commercial diets are highly digestible, absorbable, and nutritionally dense. Care should be taken to make sure patients are adequately hydrated when feeding reconstituted powdered diets; dehydration of the product within the gastrointestinal tract can lead to obstruction. This complication can be easily reversed once adequate hydration is restored.[1,4]

Determining feeding volume and frequency can be challenging for reptile patients. Caloric requirements are not well understood in reptiles, and volume tends to be the limiting factor when assist-feeding. Because of this, feeding by volume versus by caloric need tends to be more practical for clinicians.[1,4] As a general guideline, feed approximately 3% of body weight and adjust this volume based on the individual patient's needs.[9] However, chronically anorexic reptiles should be slowly returned to full feeding volumes. After initially focusing on rehydration, limit intake to only 10–15% of the daily calculated requirements and closely monitor plasma levels of phosphorus, potassium, and magnesium to prevent refeeding syndrome.[1]

Appropriate antibiotic therapy should be initiated when indicated. Culture and sensitivity testing should inform antibiotic selection when possible. In the interim, initial antibiotic therapy should be selected since most isolates in reptile patients are gram negative. A practitioner should always take the patient's health status into account and avoid medications that may be contraindicated based on any concurrent pathology. In addition, the route of administration must also be considered when selecting antimicrobials, as both ease of administration and skill level of the administrator affect client compliance. Good hydration should be maintained throughout antibiotic therapy. Common first-choice antibiotics include amikacin, ceftazidime, enrofloxacin, and trimethoprim-sulfas.[1,5]

Cardiopulmonary-Cerebral Resuscitation (CPCR)

The basic principles of CPCR apply to reptile patients despite the lack of established guidelines for these species. The airway should be secured with endotracheal intubation so oxygen therapy can be provided with positive-pressure ventilation at 6–8 breaths per minute. Cardiac compressions should be initiated in absence of a heartbeat. Administration of emergency medications (epinephrine [1:1000] 0.5 ml/kg IV, IO, or IC and atropine 0.01–0.04 mg/kg IV, IO, or IM; 0.2 mg/kg SC or IM) and IV or IO fluids to address hypovolemia should follow.[1]

Common Reptile Emergency Presentations

Traumatic

Traumatic injuries are probably the most common reason for emergency presentation in reptiles and include burns, animal bites, blunt force trauma, and fractures.[2,6,7] Ocular trauma is also quite common in reptiles and may involve either the globe itself or the spectacle.[2] As with mammals, prompt treatment is encouraged, as time can result in decreased tissue viability and increased morbidity.[2,6,7] Thermal burns are often associated with animals coming into direct contact with enclosure heating devices such as heat rocks or heat lamps.[1,2,6,7] Not all burns are immediately obvious, as reptiles do not develop the flush seen with mammals.[7] Prompt treatment and good wound management are important. Antibiotic therapy is warranted as secondary bacterial infections are common. Analgesia should be provided.[1,2,6,7]

Animal bites are extremely common. Rodent bites are often seen in snakes fed live prey items and can range from a single bite to extensive trauma if the rodent was left unattended with the snake for a prolonged period of time. Just as with burns, treatment of these injuries proceeds with wound management, antibiotics, and analgesia.[2,6,7]

Dog bites are extremely common in chelonians, and extensive shell trauma can occur. Other causes of shell trauma include lawn equipment and vehicular collisions. Grading the injury can help in development of a treatment plan and providing the owner with a prognosis. In general, the larger and deeper the injury, the graver the prognosis. Shell trauma should be treated with open wound management instead of closure with a patch, if at all possible. While patching a shell may at first seem ideal, patches preclude adequate monitoring of the wound. If a patch is to be employed, apply it only after thorough disinfection of the wound. For contaminated wounds, disinfection may mean daily lavage with sterile saline for 3–4 days. Patches should be removed after 1 month and then every 1–3 months until the wound heals, to allow for monitoring.[2,6,8]

Fractures are not immediately life-threatening and should not be considered a true emergency.[7] Fractures in reptiles fall into two types: traumatic versus pathologic. Pathologic fractures are unfortunately seen far too often in reptile patients secondary to inadequate husbandry. Fracture stabilization through application of a splint should be employed in patients exhibiting normal bone density on radiographs. Conversely, splinting can be counterproductive in patients with pathologic fracture, as application of the splint can cause further fractures to the limbs. These patients are best confined with strict cage rest. Analgesia should be provided in both cases and any husbandry issues corrected.[1,2,7]

Reproductive/Urogenital Emergencies

Dystocia is relatively common in captive reptiles and often occurs secondary to poor husbandry, but it can occur even with proper care. Problems may arise at any part of the female reproductive cycle, resulting in preovulatory, postovulatory, obstructive, or nonobstructive dystocia. Unlike with mammals and birds, emergency treatment is rarely necessary. Medical management can be pursued initially, and surgical intervention can be performed if that fails.[1,2,6,7]

As stated above, dystocia is not considered a true emergency in reptiles. The exception is if it leads to cloacal prolapse. Any cloacal prolapse—reproductive or otherwise—is a medical emergency. Structures often involved in cloacal prolapse include the bladder, phallus, oviducts, colon, or the cloaca itself. Prolapsed tissue must be identified, tissue viability assessed, and possible underlying cause(s) of the prolapse determined at presentation. Anesthesia and analgesia should be provided, and antimicrobial therapy is warranted. Viable tissue should be kept moist with a water-based lubricant until replacement is possible. Sugar, dextrose, or honey can be employed to address tissue edema to aid in prolapse reduction. Placement of stay sutures or a pursestring is recommended to prevent immediate re-prolapse.[1,2,6,7]

Nonviable prolapses should be assessed as to the feasibility of resection, especially when considering colonic and bladder prolapses. Devitalized prolapses of the phallus can be amputated without concern for urinary disruption in reptiles, as these structures are not involved in urination. Ovarian prolapse usually suggests significant reproductive pathology, and ovariosalpingectomy is recommended.[1,2,6,7]

Gastrointestinal

Gastrointestinal disease in reptiles is well described, and symptoms include regurgitation, vomiting, constipation, diarrhea, and inappetence. When gastrointestinal disease is suspected, a good place to start is with a microscopic fecal examination to rule out parasitic causes. Both a fecal flotation and a direct wet-mount cytology should be performed on fresh fecal material (<2 hours). Fecal cytology with Gram stain may also be of benefit to evaluate bacterial populations.[1,9,10] Culture can be useful, but caution is recommended when interpreting these results due to the presence of a large amount of normal flora within the alimentary canal. Additional diagnostics for the practitioner to consider include diagnostic imaging and laboratory blood testing.[18] When available, more advanced diagnostics such as magnetic resonance imaging and computed tomography may also be of use in select cases.[1]

Diagnostic imaging modalities such as radiology and ultrasonography are often beneficial in diagnosing causes of gastrointestinal disease in veterinary patients. Overall, the reptile digestive system is more difficult to visualize on survey radiographs than in mammalian patients. This difference is due to reptiles having relatively low amounts of internal fat and the relatively close apposition of organs. Regardless, the dorsoventral view will provide the best visualization of the alimentary system in chelonians and lizards. The lateral view is preferred in snakes. Some amount of intestinal gas is common in lizards, but less so in snakes and chelonians. Radiopaque foreign material may be identifiable, or enlargement of the alimentary canal can be a sign of an obstructive process. However, a prominent gas obstructive pattern is not always seen, and enlargement may also occur due to functional obstruction from ileus. Contrast procedures may be required to document occult cases of obstruction or foreign bodies. The prolonged gastrointestinal transit time documented in reptiles should be taken into account, as it may take several days for contrast agents to move through the alimentary tract.[1] Ultrasonography can be used to visualize the gastrointestinal tract of reptiles in which it can be a useful tool in detection of radiolucent foreign bodies or impacted ingesta. Changes in motility can also be observed. Intestinal wall layers can be evaluated in larger patients with the proper frequency transducer. However, large amounts of gas may interfere with this imaging modality, and overall visualization may be extremely challenging with smaller reptile species.[1]

While findings on laboratory blood analyses are rarely directly indicative of gastrointestinal disease, they can provide valuable insight. Leukograms on patients with gastrointestinal inflammation are typically characterized with heterophilia and monocytosis. Anemia may be seen in chronic cases as well. Biochemical abnormalities can be seen with gastrointestinal disease as well. Frequent changes seen in patients with vomiting, regurgitation, or diarrhea include decreases in blood protein and electrolyte levels. Elevations in aspartate aminotransferase (AST) and creatine phosphokinase (CPK) may also be seen associated with inflammation or damage to the gastrointestinal smooth muscle.[1]

Etiologies resulting in pathology of the alimentary tract include degenerative disease, gastrointestinal obstruction, chronic dehydration, metabolic derangement, suboptimal nutrition, neoplasia, trauma, toxicosis, viral or bacterial infection, and parasitism. Few degenerative gastrointestinal conditions have been described in reptiles and typically describe an atrophic process of some degree. Atrophic gastritis has been described in both the Hermann's tortoise (*Testudo hermanni*) and red-eared sliders (*Trachemys elegans*). Both cases were characterized by atrophy of the glands within the stomach mucosa, and all clinical signs were related to anorexia. Renal disease can result in mineralization of the gastric muscular tunics subsequently leading to edema, inflammation, and glandular atrophy. Atrophy can also occur in animals during prolonged periods between feedings; re-feeding leads to resolution.[11,12]

Gastrointestinal obstruction is well documented in reptile patients and can occur from ingestion of foreign material, constipation, neoplasia, or parasitic impaction. Parasitic impaction is relatively uncommon, but it can occur in cases of extremely high parasite loads. Foreign body ingestion is fairly common in reptile patients, especially the ingestion of inappropriate substrates.[11,12] Herbivorous chelonians have also been documented as ingesting woody objects with some frequency.[9] Constipation is commonly seen in those with reduced motility, functional obstructions, or chronic dehydration. It may also be seen with cystic calculi and renal hyperplasia, where either results in compression of the caudal large intestine or rectum. Both gastrointestinal neoplasia and extra-alimentary masses can result in gastrointestinal obstruction by way of either direct blockage or compression of the intestinal lumen. Anorexia and constipation are seen in females secondary to folliculogenesis as the enlarged reproductive tract decreases the space within the coelom, causing compression on the alimentary tract.[11] Intestinal intussusception has been reported in iguanas (*Iguana iguana*), chameleons (*Chameleon* spp.), and blue-tongued skinks (*Tiliqua scincoides*).[11,12] A 360° volvulus causing obstruction was documented in a hawksbill turtle (*Eretmochelys imbricate*).[12] Diagnosis of gastrointestinal obstruction relies on imaging such as radiography. Treatment of obstruction relies on resolution of the primary cause by way of either surgical correction or supportive care and medical management for partial obstructions and primary constipation. Fecal impaction that does not resolve with medical management alone requires surgical intervention.[11]

Primary gastrointestinal neoplasia has been reported in all different species of reptiles, and it can be challenging to diagnose even with extensive diagnostic testing. Associated clinical signs are nonspecific and may include poor growth, anorexia, vomiting, coelomic distention, melena or hematemesis, constipation, dysphagia, ptyalism, and palpable masses. Radiography and ultrasound can aid in diagnosis. However, definitive diagnosis requires either cytology via fine-needle aspirate or surgical biopsy.[1] Common neoplasias of the alimentary tract include adenomas, adenocarcinomas, carcinomas, papillomas, leiomyosarcomas, and sarcomas.[11] Size, location, and type of neoplasia determine treatment options that may include surgical removal, radiation therapy, and chemotherapy.[9]

The most common nutritional cause of gastrointestinal disease in reptiles is found as a sequela to nutritional secondary hyperparathyroidism. Hypocalcemia causes reduced gastrointestinal motility resulting in ileus, gaseous distension, anorexia, constipation, and colonic prolapse in some instances. Resolution of gastrointestinal signs typically occurs with resolution of the hypocalcemia. Inappropriate diets may also result in gastrointestinal pathology. Reptiles may develop diarrhea when fed diets too low in fiber. This disorder is most commonly documented in herbivorous reptiles when low-fiber diets cause disruption of the fermentation cycle, bloating, and dysbiosis resulting in lactate-induced diarrhea. Dietary correction often leads to resolution of clinical signs.[11]

Gastrointestinal toxicosis is infrequently seen reptile patients. Lead toxicosis has been documented in Greek tortoises (*Testudo graeca*) and alligators after ingestion of lead shot. Ingestion of toxic invertebrates, such as fireflies (*Photinus* spp.), are also well documented in reptiles. Gastritis may also occur iatrogenically after treatment with nonsteroidal anti-inflammatory medications or corticosteroids. As with toxicosis in other species, treatment is specific to the toxin ingested and coupled with appropriate supportive care.[1]

Various infectious agents can result in gastrointestinal disease in reptiles. Bacterial gastroenteritis can occur secondary to disruption of the normal flora that forms a protective barrier within the alimentary canal. Factors such as inappropriate diet, stress, antimicrobials, and any disease disrupting absorption of nutrients play a key role. Common isolates are primarily gram-negative, such as *Escherichia coli*, *Pseudomonas* spp., *Serratia* spp., *Aeromonas* spp., *Citrobacter* spp., *Pasteurella* spp., *Alcaligenes* spp., *Proteus* spp., and *Klebsiella* spp. However, other organisms have been isolated, such as *Clostridium* spp., *Mycobacteria* spp., *Staphylococcus* spp., and *Corynebacterium* spp.[1,11] For example, *Mycobacteria* spp. has been reported causing stomatitis in Boa constrictors (*Boa constrictor*).[1] Diagnosis is made by microscopic examination of fecal samples and fecal culture. Fecal cytology should be performed by way of Gram and acid-fast staining. If a clostridial etiology is suspected, toxin identification is recommended. Treatment involves hydration, supportive therapy, dietary optimization, and broad-spectrum antibiotic therapy based on culture and sensitivity results.[1,11]

Adenoviral enteritis has also been documented in crocodilians and rosy boas (*Lichanura trivirgata*).[1,11] Adenoviral infections have been identified in other species of reptile with gastrointestinal clinical signs, yet it is not always clear whether it is occurring as the primary pathogen. Reovirus was associated with acute necrotizing colitis in a group of six juvenile Arizona mountain kingsnakes (*Lampropeltis pyromelana*) that responded to supportive care and antibiotic treatment for secondary bacterial invasion. Herpesvirus was found to be the cause of illness in two monitor lizards (*Varanus* spp.) with multifocal necrosis of the lamina propria of the small intestine and liver.[1,12] Inclusion body disease (IBD) in snakes, a disease recently confirmed to be caused by an arenavirus, has been long documented as a cause of chronic regurgitation, anorexia, and weight loss usually accompanied by the hallmark neurologic disease. Stomatitis is also seen with some frequency in IBD-positive snakes.[1] No specific treatment exists at this time specifically for treatment of viral gastrointestinal disease.[1,12]

Gastrointestinal parasitism is the most common etiology observed in clinical reptile practice.[14] It is estimated that virtually 100% of wild reptiles harbor at least one kind of parasite.[15] The presence of a parasite alone does not necessarily result in clinical disease.[15,16] Prior to determining that an organism is truly parasitic, one must establish that the organism is known to cause pathogenicity and that it is causing damage in some way to the host, either directly or by interfering with various metabolic functions.[16] Other factors contributing to pathology include stress, inappropriate environmental temperatures, hygiene, concurrent disease, parasitic load, nutritional status, and age.[15,16] Superinfection—marked increases in parasite population density—can play an important role in parasite virulence. Ancillary diagnostics, such as complete blood count and blood biochemical analysis, can be a useful tool in assessing virulence.[16] Parasites with direct lifecycles tend to be more of a concern in clinical practice due to lack of availability of intermediate hosts for parasites with indirect lifecycles.[15] Clinical signs associated with parasitic infestation arise from the parasite directly competing for nutrients, removing tissue or blood from the host, vascular or lymph obstruction, tissue damage and ulceration, tissue necrosis, inflammation, and edema. Few parasitic diseases cause pathognomonic clinical signs. Definitive diagnosis of parasitism is made by identification of the organism either by direct visualization, microscopy, or detection with polymerase chain reaction analysis.[15,16]

Neurologic

Common neurologic emergencies in reptile patients include acute ataxia, paresis or paralysis, seizures, or muscle fasciculations. As with mammalian patients, laboratory blood analysis and imaging should follow a thorough history and physical examination. The most common causes of neuropathies in reptiles include nutritional secondary hyperparathyroidism, intoxications, trauma, end-stage renal disease, and envenomations. Treatment is dependent on the underlying cause and severity of the condition.[1]

Respiratory

Most respiratory emergencies in reptiles are the result of chronic, often untreated illness. True acute respiratory emergencies are rare and result from airway obstruction and drownings.[1] Dyspnea needs to be differentiated from panting or behavioral posturing.[2] Dyspnea can be primary or secondary to pulmonary compression from intracoelomic fluid or masses. Compensatory respiratory alkalosis is also seen in reptiles suffering from severe metabolic derangements.[1] Blood testing and imaging should be utilized to determine the underlying cause.[2] Oxygen supplementation should be provided when necessary. Drowning patients presenting in respiratory arrest often will respond to CPCR and positive-pressure ventilation following drainage of fluid.[1]

Other Emergency Presentations

As discussed earlier, it is not uncommon for reptiles to present on an emergency basis with manifestations of chronic illness, such as septicemia or renal failure. These animals may present obtunded to moribund with varying degrees of weight loss. Diagnosis and prognosis should be determined with laboratory testing and imaging, with appropriate treatment to follow.[1,2]

Conclusion

Reptiles present on emergency for a number of reasons, whether a true emergency or perceived as one by the owner. While acute emergencies are not commonplace in reptile medicine, it is important to be prepared for them. A thorough understanding of reptiles as patients coupled with proper patient assessment allows the practitioner to best utilize available therapeutics to result in the most favorable outcomes possible.

References

1. Divers SJ, Stahl SJ. *Mader's Reptile and Amphibian Medicine and Surgery*. 3rd ed. St. Louis, MO: Saunders Elsevier; 2019.

2. Music KM, Strunk A. Reptile critical care and common emergencies. *Vet Clin Exot Anim*. 2016;19:591–612.

3. Cambell TW. Clinical pathology of reptiles. In: Divers S, Mader DR, eds. *Reptile Medicine and Surgery*. 2nd ed. St. Louis, MO: Saunders Elsevier; 2006.

4. De Voe RS. Nutritional support of reptile patients. *Vet Clin Exot Anim Pract*. 2014;17:249–261.

5. Mader DR. Antibiotic therapy in reptiles. In: Proceedings of the Central Veterinary Conference. 2008. www.dvm360.com/view/antibiotic-therapy-reptiles-proceedings-0 (VIN editor: Original link was modified as of 8-3-2020).

6. Mader DR. Common emergencies in reptiles. In: Proceedings of the Fetch DVM360 Conference. 2008.

7. Klaphake E. Managing the top five most common reptile emergencies. In: Proceedings of the NAVC Conference. 2013.

8. Bjornebo HA. Treatment of traumatic shell injuries in turtles and tortoises. VIN/ARAV Rounds. Nov. 17, 2019. beta.vin.com/members/cms/project/defaultadv1.aspx?id=9395371&pid=24494&.

9. Nevarez J. Lizards. In: Mitchell MA, Tully TN, eds. *Manual of Exotic Pet Practice*. St. Louis, MO: Saunders Elsevier; 2009:164–206.

10. Kirchgessner M, Mitchel MA. Chelonians. In: Mitchell MA, Tully TN, eds. *Manual of Exotic Pet Practice*. St. Louis, MO: Saunders Elsevier; 2009:207–250.

11. Benson KG. Reptilian gastrointestinal diseases. *Semin Avian Exot Pet Med*. 1999;8(2):90–97.

12. Reavill DR. Pathology of reptilian gastrointestinal tract. In: Proceedings of the Annual Conference of the Association of Reptilian and Amphibian Veterinarians. 2011:72–88.

13. Hetzel U, Sironen T, Laurinmaki P, *et al*. Isolation, identification, and characterization of novel arenaviruses, the etiological agents of boid inclusion body disease. *J Virol*. 2013;87(20):10918–10935.

14. Scullion FT, Scullion MG. Gastrointestinal protozoal diseases in reptiles. *J Exot Pet Med*. 2009;18(2):266–278.

15. Wilson SC, Carpenter JW. Endoparasitic diseases of reptiles. *Semin Avian Exot Pet Med*. 1996;5(2):64–74.

16. Maas III AK. Protozoal parasitology for the reptile practitioner. In: Proceedings of ExoticsCon. 2016:45–47.

Speaker Information
(click the speaker's name to view other papers and abstracts submitted by this speaker)

Heather A. Bjornebo, DVM, DABVP (Reptile & Amphibian Practice), CertAqV
Arizona Exotic Animal Hospital
Mesa, AZ, USA

URL: https://www.vin.com/doc/?id=9768904

Reptile Behavior, Enrichment, and Training

ExoticsCon Virtual 2020 Proceedings

Anna-Marie Ruoff, DVM
Arizona Exotic Animal Hospital, Mesa, AZ, USA

Abstract

Reptiles have traditionally been considered unthinking and unfeeling animals. However, recent research has revealed complex intelligence, cognition, and behaviors in various reptile species. Many owners, keepers, and veterinarians are now considering and discussing the mental well-being of reptiles in captivity—including providing enrichment through diet, husbandry, and behavioral training. Enrichment benefits reptiles by encouraging natural behaviors, increasing choice, and enhancing welfare. Behavioral training can be utilized to provide enrichment, reduce stress, facilitate veterinary care, and improve human-animal interactions. This masterclass lecture will include reviews of reptile intelligence and cognition, unique reptile behaviors, enrichment opportunities in captivity, and behavioral training of reptiles.

Introduction

Reptiles have traditionally been considered unthinking and unfeeling animals.[1] However, recent research has revealed complex intelligence, cognition, and behaviors in various reptile species.[1] Coverage of this research in popular science, media, and herpetology magazines has increased public appreciation for reptile behavior. As the popularity of reptile pets grows, there is increased interest in improving their care.[1-3] Many owners, keepers, and veterinarians are now considering and discussing the mental well-being of reptiles in captivity—including providing enrichment through diet, husbandry, and behavioral training. Enrichment benefits reptiles by encouraging natural behaviors, increasing choice, improving health, and enhancing welfare.[1,4] Behavioral training can be utilized to provide enrichment, reduce stress, facilitate veterinary care, and improve human-animal interactions.[4]

Veterinarians play an integral role in improving the health and welfare of captive reptiles.[5] Failure to provide for the physical needs of reptiles is a source of stress and is recognized as a significant cause of health problems.[6] Physical and mental health are intricately interconnected; veterinarians must consider and address both in order to provide the best medical care for their patients and enhance the lives of captive reptiles.[5,6]

Behavior

Each of the more than 10,000 known species of reptiles has a unique natural history, including varied diets, environments, and behaviors.[7] It is important for those who keep and care for reptiles—including veterinarians—to become familiar with the natural histories of the species they manage.[2] Knowledge of a specific species' natural history can inform choices for their captive care, including diet, husbandry, enrichment, and medical evaluation.[2,8]

Ethology is the study of behavior from a naturalistic and evolutionary perspective.[1] It requires careful description of behaviors observed in natural and captive conditions, including the behavior itself as well as its stimulus or context.[1] Reports of behaviors are typically organized systematically, including locomotion, foraging and ingestion, body care and elimination, protection against predators, fighting and agonism, territory formation and maintenance, dominance, courtship, nesting, and parental care.[1]

Historically, behavioral research neglected reptiles relative to birds and mammals, especially with regard to cognition, emotions, social behaviors, and enrichment needs.[1] Much of what we know about reptile behavior is from sparse primary research and informal or anecdotal reports that lack thorough documentation.[1] However, reptiles share many traits with mammals and birds, including complex communication, problem-solving, parental care, play, and social behaviors.[1,9] Reptile behavior has historically been interpreted through a mammal-centric perspective. Burghardt explains, "We cannot use simplistic evaluations of behavior viewed through our human-centered lens, since poikilothermic vertebrates operate at different time scales, lack readily interpreted facial and vocal signals of internal states, and rely on sensory cues, often chemical, outside the range of those we can readily evaluate with our mammalian infrastructure."[1]

The diversity of behaviors among reptile species is comparable to mammals and birds.[8] Within the reptile clade, there are large variations in sensory abilities, temperaments, reproductive strategies, neonatal strategies, anti-predator responses, foraging strategies, social organization, and habitat selections.[1]

Reptile senses include visual, auditory, tactile, and chemosensory systems. Most reptiles have good vision, which is used in foraging and predator avoidance, as well as social behaviors including head bobbing, body coloration changes, courtship gestures, and submissive postures.[3,7] Hearing is relatively well developed in reptiles; many reptiles vocalize and/or respond to auditory stimulation.[3,7] Tactile sensation is evident in the detection of vibrations, as well as in behaviors such as stroking or rubbing in courtship, or ramming or biting in fighting.[3] Chemosensation is also well developed and is used in communication, reproduction, foraging, and predator avoidance.[3,7] Consideration for these sensory systems is important in captive husbandry, including in identifying sources of stress as well as providing enrichment.

Most reptile species are not considered social, but there are numerous reports of social behavior in various species.[1,6,9] Some species have social dynamics that change with life stage or reproductive status.[6] Examples of social behavior include courtship, territoriality, monogamy, social grouping, definite social orders, and parental care.[1,2,6,7,9]

Intelligence and Cognition

Reptiles were historically considered to be unthinking and unfeeling—their behaviors were interpreted as mere responses to stimuli without higher cognition.[1] Indeed, colloquial references to the "lizard brain" belie their more complex neurologic anatomy; some reptiles have neural structures homologous to structures in mammals that govern emotion, decision-making, and consciousness.[1,10] There have been relatively few studies about reptile intelligence and cognition compared to mammals and birds.[1] Early studies also did not take differences between reptiles and mammals into account when establishing measures of intelligence.[1]

Reptiles have been shown to have signs of intelligence and cognition, including spatial cognition, sensory discrimination, reversal learning, equivalence, individual recognition, gaze following, complex social dynamics, social learning, curiosity and play behaviors, and problem-solving.[1,9] Examples of behaviors supportive of cognition include counting-like skills in monitors, play behaviors in Komodo dragons and turtles, and predator gaze monitoring in hognose snakes feigning death.[1]

Reptiles also have the abilities to learn and be trained when the training accommodates their sensory systems and behavioral repertoires.[1,3,4,6,9] Burghardt argues that lack of accommodation to human training does not necessarily reflect an animal's lack of intelligence.[1] Turtles, crocodilians, and lizards are especially adept with traditional learning and training techniques.[1] Some turtles have demonstrated prolonged memories, remembering how to complete a training task years after reinforcement last occurred.[1] Snakes may rely more on chemosensory cues that are less familiar to their human trainers[1]; however, indigo snakes were shown to respond to operant conditioning at a similar rate as pigeons or rats.[3] More research into snakes' receptivity to training is necessary.[1]

Stress

Stress or stressors are any physical, chemical, or emotional force that disturbs or threatens homeostasis and the accompanying adaptive responses (the stress response) that attempts to restore homeostasis.[5,6] Stressors are categorized as physical (those necessitating physiologic coping mechanisms) or psychologic (those that may indicate imminent threat but are not directly harmful).[5,6] Physical stressors include inadequate or excessive environmental temperature, humidity, or photoperiod; thirst or dehydration; hunger or undernutrition; inappropriate diet or malnutrition; oxygen deprivation; disease; trauma; and reproduction.[5,6] Psychologic stressors include social interactions, overcrowding, confinement, novel stimuli or environments, human-reptile interactions, and lack of appropriate seclusion.[5,6]

Stress responses can be categorized by their chronicity. Responses to short-term stressors are typically adaptive measures to increase energy for the animal to cope with or escape the stressor. Short-term stress responses include activation of the sympathetic nervous system, the hypothalamic-pituitary-adrenal axis, and other endocrine functions.[5,6] Responses to chronic stressors may become maladaptive through dysregulation, often due to the inability of the animal to cope with or escape the chronic stressor.[5,6] Chronic stress can lead to the inhibition of non-essential behaviors, reduction of physiologic functions, and immunosuppression.[2,5,6]

A common cause of chronic stress is long-term husbandry deficiency.[5,6] Captivity is inherently stressful for non-domesticated species, because it cannot fully provide for all of an animal's needs.[1,2,6,8] Reptile keepers' attempts to mimic species' natural environments are based on limited knowledge and abilities.[1,8] For example, enclosures are often a fraction of the size of normal territories, and artificial lighting is a poor alternative to the sun. Captivity also affects behavior. Behaviors exhibited in captivity, such as escape attempts or nose rubbing, can indicate distress.[6,7] More subtle signs of stress may be overlooked or misinterpreted by keepers, but can include abnormal behaviors, behavioral inhibition, hypervigilance, excessive hiding, startle, aggression, or immobilization.[6] Reptiles may change their behavior in the presence of human observers in captivity, and they also often mask signs of disease.[1,5,6,8]

Relevance to Veterinary Medicine

Evaluating a reptile's behaviors is an important part of its clinical assessment.[5-8] Physical and mental health are interconnected; physical health will influence behaviors, and mental health will influence physiologic responses.[5,6] Behavioral change in reptiles is often an early indicator of distress, injury, or disease.[8] Stressed reptiles may display false indications of physical illness, such as erratic behavior, feigning death, or increased respiration rates.[3] Interpreting subtle behavioral changes necessitates familiarity with the species' natural history and behaviors, as well as the patient's individual history.[5] An unfamiliar keeper or clinician may misinterpret normal behaviors as indicators of disease, or abnormal behaviors as normal.[7]

Maladaptive or abnormal behaviors may result in physical injury or disease, such as rostral abrasions, thermal burns, injuries due to conspecific aggression, or intestinal impactions due to pica.[2,8] Chronic stress is detrimental to health and welfare, including immune function, reproduction, and growth.[2,5,6] Medical management of reptiles must incorporate both treatment of the pathologic process as well as identifying and reducing stressors in the environment.[6]

There are also iatrogenic stressors inherent with medical care, including transportation, restraint, and diagnostic procedures.[6] Clinicians must balance the potential benefit of a procedure against its potential for causing stress.[6] Stress responses may also affect results of the physical examination and biochemical laboratory results.[3] Stressing a compromised animal with restraint may result in severe harm or even death.[3] Strategies for minimizing stress should be utilized during the examination, including appropriate environmental conditions, reduced stimuli, examining at a distance when possible, and behavioral training.[3]

Enrichment

Enrichment is a popular and effective strategy for enhancing captive animal welfare.[1,6] The Association of Zoos and Aquariums Behavior Scientific Advisory Group defines enrichment as "a dynamic process for enhancing animal environments within the context of the animal's behavioral biology and natural history...with the goal of increasing the animal's behavioral choices and drawing out their species-appropriate behaviors, thus enhancing animal welfare."[6] The purpose of enrichment is to create an environment that enables an animal to have opportunities to exhibit natural behaviors similar to a wild conspecific.[1,2,4,6] Enrichment can also be used to reduce or eliminate chronic stress.[2,5] Enrichment is sometimes considered a luxury that keepers may choose to provide, but it actually may be essential for proper captive management.[1] Good enrichment should be based on knowledge of the ethology and natural history of a species, and it should encompass all aspects of husbandry.[1,4]

A survey of reptile zookeepers indicated the main goals of their enrichment programs included promotion of natural behaviors, behavioral conditioning, reducing stereotypies, providing public education, and the promotion of general well-being.[11] More broadly, the goals of an enrichment program should promote species-appropriate behaviors, provide behavioral opportunities, and provide animals with choices or control over their environment.[4] The incorporation of choice and control can help modulate stress response in many species.[5]

When designing an enrichment program, one should have a specific and measurable goal in order to evaluate the success of the program.[4] Examples of goals include utilization of the provided enrichment, increase in a desired behavior, reduction of an undesired behavior, improved biological growth measurements and stress indicators, or decreased corticosterone levels.[11] The "SPIDER" framework used by some zoos includes the following steps: Setting goals, Planning, Implementing, Documenting, Evaluating, and Readjusting.[4,12]

"Opportunities to Thrive" is another framework for improvements in animal care and welfare created by San Diego Zoo Global.[5,6] The five opportunities are for a well-balanced diet (including varied presentation), to self-maintain (including shelter and comfort behaviors), for optimal health, to express species-specific behaviors, and for choice and control (the choice to perform a variety of behaviors or not).[5,6]

When designing enriching enclosures, considerations must be made for the species' natural history and ethology, although natural conditions are difficult to approximate in captivity.[1,2,6,8] Methods for providing enrichment include natural enrichment devices, manmade enrichment devices, social enrichment, and structure or habitat design.[11] Considerations for enclosures include temperature ranges, thermoregulation behaviors, activity levels, vertical and horizontal space, substrate type and depth, airflow, olfaction or chemosensory cues, lighting spectrum, photoperiod, humidity, drinking water provision, variety of diet, foraging or hunting behaviors, diet presentation, shelter types, anti-predator behaviors, visual barriers (including tolerance for "invisible" barriers such as glass), social behaviors, and crowding.[2,5,6] Choice is also an important consideration; many of the variables listed above should be provided on a gradient to allow the reptile to choose its microhabitat, including temperature, lighting spectrum, humidity, and shelter.[2] Temporal variation should also be considered for species whose natural history includes seasonal variation.[2] Naturalistic environments provide more enrichment and less stress than clinical environments.[2,6] However, naturalistic enclosures may be more difficult to maintain and clean than clinical enclosures.[2,6] Enriching enclosures may be designed without naturalistic aesthetics by utilizing artificial material to accomplish the same enrichment goals. Frequent changes to enclosures can provide novelty and enrichment; however, frequent changes to an animal's environment may also be a source of stress.[5] The tolerance of the individual animal should also be considered when designing enrichment.[5]

Relevance to Veterinary Medicine

Veterinarians should encourage keepers to improve their reptiles' physical and mental health by providing species-appropriate enrichment, including choices and opportunities to express natural behaviors.[5] Enrichment can enhance health and welfare, decrease stress (especially chronic stress), and improve an animal's ability to cope with stressors.[3] This author includes discussions of enrichment along with husbandry and diet requirements during routine and medical examinations of reptiles.

Training

Behavioral training is one method of enrichment. Training benefits the reptile by providing enrichment, physical activity and exercise, mental stimulation, improved welfare, and reduced stress associated with human interaction.[2,3] Training can also benefit the human trainer by providing increased interaction with the reptile, increased safety during interactions, ease of movement of large reptiles, reduced necessity of restraint for medical procedures, and potential information for research in animal intelligence.[3,4] Many reptile owners want to be able to interact more meaningfully with their pet reptile; training provides a method of communicating desired actions and promised rewards, and is enriching for human-reptile interaction.[4]

Operant conditioning is commonly studied in mammals and birds. While fewer published studies describe operant conditioning in reptiles, it is commonly used in zoological institutions and by private keepers.[4] Many techniques for training in reptiles have been extrapolated from those used with mammals and birds.[3] A survey of reptile zookeepers found that their sources for reptile enrichment and training ideas included word of mouth from other keepers, the internet, peer-reviewed published journal articles, zoo-accreditation-supplied networking, and herpetology magazines.[11] There are also social media groups dedicated to reptile training, which may provide ideas and networking for reptile keepers but are not peer-reviewed published techniques. Generally, operant conditioning should rely on positive reinforcement as much as possible to reduce fear response.[3]

When establishing a training program, the trainer must recognize differences between mammals or birds and reptiles, including sensory capabilities, social dynamics, and metabolic rates. Reptiles that lack external ears may not respond to auditory cues, but a variety of visual, auditory, tactile, and/or olfactory cues should be considered.[3] Reinforcements are typically dietary, but these are not necessarily appropriate for species that feed infrequently (such as many snakes) or species that are prone to obesity in captivity.[3] Determining the best reinforcement for an individual animal takes careful observation to recognize what the animal values.[5] Trainer safety must also be considered, as some reptiles can become aggressive when anticipating food; the reinforcement can be delivered with tongs or similar equipment.[3]

A relatively simple form of training is desensitization, which is the process of acclimating an animal used to a stimulus through gradual exposure.[3] Desensitization may be passive (habituation) or active (counterconditioning). Habituation involves introduction of the stimulus in a way that allows the animal to choose to interact with it, but does not force the interaction.[3] Counterconditioning is an active pairing of the stimulus with a reinforcement to create a positive association; it is generally faster than habituation.[3] Desensitization is a useful technique for getting a reptile used to human handling.[2,3,6] It is important to provide a cue before handling to reduce the startle response, since many reptiles are prey species.[3]

Target training can be accomplished in almost any animal species.[3,5] It is a very powerful training tool that can be used as a bridge to train other behaviors.[1] A target is an object that an animal is taught to touch, follow, or approach; common targets include ball on a stick, human hand, straw, spoon, or dish.[3] A positive association with the target is formed through operant conditioning. Target training can also be used to teach stationing, which is to direct an animal to go to a specific item or location and remain there.[3] Stationing is useful for shifting animals into and out of transport cages and reducing physical or chemical restraint necessary for medical procedures.[2,3] Target and station training may provide safe methods to position and manipulate even large or dangerous reptiles.[3] Stations can be brought to novel environments such as veterinary offices to provide a consistent training framework and reduce stress.[3]

Relevance to Veterinary Medicine

Training techniques may facilitate veterinary care and reduce associated stress.[2-4] Decreased stress can increase the efficacy of a procedure and validity of diagnostic results.[3]

Target and station training can reduce stress associated with capture and transportation for both the reptile and the keeper.[2,4] Reptiles that are desensitized to human handling are less likely to experience stress when handled during medical procedures.[3,5] Desensitization training should incorporate the veterinarian, touch, equipment, and common medical procedures.[3] Reptiles may be desensitized to diagnostic procedures such as weighing, ultrasounds, or venipuncture.[3] Notably, Aldabra tortoises were trained to station for venipuncture.[13] Target, station, and desensitization training can also improve safety for the handler and animal by reducing fear behaviors and resultant injuries.[3]

Additionally, keepers who have routine interaction with their reptiles may be more aware of their normal and abnormal behaviors.[3] These keepers may notice earlier or more subtle signs of disease in the reptile, which would allow for veterinary care to be delivered earlier in the progression of the disease and potentially a better prognosis.[3] People who have regular or meaningful interactions with their reptile may also become more invested in its care.[3]

Conclusion

Reptiles can have complex intelligence, cognition, and behaviors. Many are capable of learning and can respond to operant conditioning and other training. Enrichment through environmental, dietary, and behavioral training modalities is important for the health and welfare of reptiles in captivity. Behavioral training can be utilized to provide enrichment, reduce stress, facilitate veterinary care, and improve human-reptile interactions. Behavior, stress, enrichment, and training are important considerations for veterinarians who treat reptile species. Continued research into reptile behavior, intelligence, enrichment, and training is important due to the relative paucity of published information. This author encourages readers interested in reptile behavior, intelligence, and training to explore the resources and references provided for more information to develop their own enrichment, training, and/or research programs.

Acknowledgements

I would like to thank Arizona Exotic Animal Hospital; Jay Johnson, DVM; and Heather Bjornebo, DVM, DABVP (Reptile and Amphibian Practice), CertAqV, for their continued support.

References

1. Burghardt GM. Environmental enrichment and cognitive complexity in reptiles and amphibians: concepts, review, and implications for captive populations. *Applied Animal Behaviour Science*. 2013;147:286–298.

2. Wilkinson SL. Reptile wellness management. *Vet Clin North Am Exot Anim Pract*. 2015;18:281–304.

3. Hellmuth H, Augustine L, Watkins B, Hope K. Using operant conditioning and desensitization to facilitate veterinary care with captive reptiles. *Vet Clin North Am Exot Anim Pract*. 2012;15:425–443.

4. Skurski ML, Fleming GJ, Daneault A, Pye GW. Behavioral training and enrichment of reptiles. In: Divers SJ, Stahl SJ, eds. *Mader's Reptile and Amphibian Medicine and Surgery*. 3rd ed. St. Louis, MO: Elsevier Saunders; 2019:100–104.

5. Clayton LA, Tynes VV. Keeping the exotic pet mentally healthy. *Vet Clin North Am Exot Anim Pract*. 2015;18:187–195.

6. Hunt CJG. Stress and welfare. In: Divers SJ, Stahl SJ, eds. *Mader's Reptile and Amphibian Medicine and Surgery*. 3rd ed. St. Louis, MO: Elsevier Saunders; 2019:105–108.

7. Durso AM, Maerz JC. Natural behavior. In: Divers SJ, Stahl SJ, eds. *Mader's Reptile and Amphibian Medicine and Surgery*. 3rd ed. St. Louis, MO: Elsevier Saunders; 2019:90–99.

8. Warwick C, Arena P, Lindley S, Jessop M, Steedman C. Assessing reptile welfare using behavioral criteria. *In Practice*. 2013;35:123–131.

9. Wilkinson A, Huber L. Cold-blooded cognition: reptilian cognitive abilities. In: Shackelford TK, Vonk J, eds. *Oxford Handbook of Comparative Evolutionary Psychology*. Oxford: Oxford University Press; 2012:129–143.

10. Norimoto H, Fenk LA, Li H, *et al*. A claustrum in reptiles and its role in slow-wave sleep. *Nature*. 2020;578:413–418.

11. Eagan T. Evaluation of enrichment for reptiles in zoos. *J Appl Anim Welf Sci*. 2019;22:69–77.

12. Animal Enrichment. Available at: www.animalenrichment.org. Accessed April 5, 2020.

13. Weiss E, Wilson S. The use of classical and operant conditioning in training Aldabra tortoises (*Geochelone gigantea*) for venipuncture and other husbandry issues. *J Appl Anim Welf Sci*. 2003;6:33–38.

SPEAKER INFORMATION
(click the speaker's name to view other papers and abstracts submitted by this speaker)

Anna-Marie Ruoff, DVM
Arizona Exotic Animal Hospital
Mesa, AZ, USA

URL: https://www.vin.com/doc/?id=9768903

AEMV

Chinchillas: Clinical Techniques and Common Disorders

ExoticsCon Virtual 2020 Proceedings

Christoph Mans, DrMedVet, DACZM, DECZM (Zoo Health Management)

School of Veterinary Medicine, University of Wisconsin-Madison, Madison, WI, USA

Introduction

Chinchillas are long-lived hystricomorphic rodents, which have been domesticated just a century ago. Chinchillas are used as animal models in otological research and are becoming increasingly popular as companion animals. Chinchillas are adapted to a dry and cold climate, and in captivity relative humidity should be less than 60%, with less than 50% being ideal. Environmental temperature should be less than 24°C (75°F). While chinchillas are very tolerant to cold temperatures, high temperatures are not tolerated, and heat stroke may occur if the environmental temperature is higher than 27°C (80°F).

Chinchillas are social animals and are ideally kept in groups. Large cages with multiple levels are preferred (e.g., ferret cages) and single-level rabbit or guinea pig cages are not appropriate. A hiding box or other hiding structures should be provided, and if several animals are housed together, sufficient hiding space should be provided. A dust (sand) bath should be provided for short periods of time on a regular basis (at least once weekly). Commercial chinchilla sand should be used. Regular dust bathing is important, in order to allow chinchillas to maintain a healthy coat.[1]

Chinchillas are obligate herbivores and are adapted to a natural diet low in water. Therefore, chinchillas should not receive fresh produce in significant amounts, in order to prevent GI disturbances. Chinchillas should be fed a diet consisting of high-quality grass hay and commercial chinchilla pellets. As treats, dehydrated vegetables (e.g., carrots) and other commercial herbivore rodent treats are suitable. Raisins and other food items high in sugar and phosphorus should not be fed in order to reduce the risk of development of dental disease. Fresh water should be provided in a bowl or bottle.[1]

Chinchillas can be handled and restrained like most other exotic companion mammals, but care should be taken to avoid "fur slip," which will occur if chinchillas are grasped by their fur.[1,2]

Diagnostic Techniques

The diagnostic techniques used in chinchillas are the same as in other species; however, some diagnostic tests are fairly unique to chinchillas or have high diagnostic value. For blood collection the author prefers insulin syringes with 27–28-gauge needles in order to maximize the chance to obtain a diagnostic blood sample. The preferred venipuncture sites are the jugular veins, which can usually not be visualized, except in cachexic animals. Jugular venipuncture can be performed safely without anesthesia or sedation and usually provides the largest blood samples. The lateral saphenous vein may provide a blood sample up to 0.2 milliliters, but often only insufficient blood samples can be collected. Other venipuncture sites include the femoral vein and ventral tail vein.

Urinalysis is another valuable diagnostic test, which is currently underutilized in chinchillas. The normal urine coloration can vary considerably. The urine pH should be ≥8.5; proteinuria is normal in chinchillas and the most common urinary crystals are amorphous, while calcium carbonate crystals are not typical.[2] Urinalysis is also helpful to detect ketonuria in anorexic chinchillas and can be used to monitor response to treatment in these animals.

Diagnostic imaging (radiography, computed tomography and ultrasonography) may require sedation or anesthesia, depending on the imaging modality and individual animal.

Several anesthetic protocols have recently been evaluated in chinchillas.[3-5] Isoflurane anesthesia has the fewest post-anesthetic effects on food intake and fecal output. Dexmedetomidine-ketamine (DM, 0.015 mg/kg + 4 mg/kg IM/SC) anesthesia has comparable cardiovascular effects to isoflurane anesthesia and is considered safe and reliable. The DM protocol is used routinely by the author for a variety of procedures, including dental procedures.

Common Disorders

Ketoacidosis and hepatic lipidosis are very common in chinchillas and occur secondary to any disease condition which results in a negative energy balance, usually due to anorexia. Urine analysis is a simple and noninvasive diagnostic tool to diagnose ketoacidosis and monitor response to treatment. Routine plasma biochemistry will not aid in the diagnosis of ketoacidosis. Treatment should focus on the primary underlying cause, which led to anorexia, while providing nutritional support and fluid therapy.[1]

Gastrointestinal disorders commonly diagnosed in chinchillas include constipation and tympany, while diarrhea is less common. Causes for constipation are insufficient food intake secondary to other underlying disorders, enteritis and metabolic disorders. Tympany occurs secondary to dysbacteriosis or hypomotility of the GI tract. Less commonly, rectal tissue prolapse or intestinal intussusception can occur. Soft feces or diarrhea are most common in younger animals. Common underlying causes include sudden diet changes, inappropriate amounts of fresh vegetables or fruits, endoparasites, or bacterial overgrowth due to inappropriate oral antibiotic use.

Dental disorders, including cheek teeth elongation, periodontal disease, tooth resorption and caries are common in chinchillas.[6,7] Periodontal disease and resorptive lesions are more common in middle-aged or older animals. The incisor and cheek teeth of chinchillas grow life-long (elodont), have no anatomic root (aradicular) and have a large crown (hypsodont). The labial aspects of the incisor teeth are physiologically orange in color. Four cheek teeth, which are morphologically identical, are present in each dental arcade. The occlusal surface is 0 degrees, and the clinical crown is very short. Excessive salivation (slobbers) is characteristic for chinchillas with intraoral pain or dental disease. Palpation of the ventral mandibular aspects may reveal boney irregularities, which occur secondary to apical cheek tooth elongation. Periodontal disease, tooth resorption and caries are very common in chinchillas and can be easily missed.

Otitis externa is uncommon in chinchillas and if present usually developed secondary to otitis media and perforation of the tympanic membrane. Therefore, diagnostic imaging of the skull should be performed in all chinchillas diagnosed with otitis externa. Otitis media is usually bacterial in origin, and *Pseudomonas aeruginosa* is commonly isolated. Sterile sampling of the middle ear for cytology and bacterial cultures can be performed by a percutaneous technique developed for chinchillas. The bulla is accessed in sterile fashion through its dorsal wall. This technique is performed under general anesthesia and is well tolerated.[6]

Ocular disorders are common in chinchillas. Epiphora (wet eye), conjunctivitis and corneal trauma and keratitis are commonly diagnosed. The large and exposed cornea and the only rudimentarily developed third eyelids may predispose chinchillas to corneal trauma. *Pseudomonas aeruginosa* is the most common isolate from chinchillas with conjunctivitis.[8] Therefore, empiric topical antibiotic treatment, pending bacterial culture and sensitivity results, should include gentamicin- or polymyxin B-containing ophthalmic formulations.

Female reproductive tract diseases are not common in pet chinchillas. Endometritis and pyometra are the most common disorders. In breeding females, fetal retention and mummification may be seen. Neoplasia of the reproductive tract in chinchillas is uncommon.

Male reproductive tract diseases are limited to the penis and prepuce. Fur rings, balanoposthitis, preputial abscesses, paraphimosis, and phimosis are diagnosed.[9]

Urolithiasis in chinchillas is uncommon and affects predominately males. In all reported cases the calculi were composed predominately of calcium carbonate.[10] Hematuria is the most common clinical sign and a single or several calculi can be found in the bladder and/or urethra.[10] Urethral calculi carry a worse prognosis compared to cystic calculi. The rate of recurrence of urolithiasis in chinchillas following surgery has been reported to be 50%.[10]

Cardiac disease is uncommon in chinchillas, while the presence of innocent (syn. physiologic) heart murmur is somewhat common. Echocardiography is recommended for evaluation of chinchillas with heart murmurs grade III/VI or higher or in animals with clinical signs likely associated with cardiac disease.[11] Vertebral heart size (VHS) reference intervals using radiographs or computed tomorgraphy have been published in chinchillas. The radiographic VHS reference interval in chinchillas is 7.5–10.2.[12]

References

1. Mans C, Donnelly TM. Disease problems of chinchillas In: Quesenberry KE,Carpenter JW, eds. *Ferrets, Rabbits, and Rodents: Clinical Medicine and Surgery.* St. Louis, MO: Elsevier Saunders; 2012:311–325.

2. Doss GA, Mans C, Houseright RA, *et al.* Urinalysis in chinchillas (*Chinchilla lanigera*). *J Am Vet Med Assoc.* 2016;248:901–907.

3. Fox L, Snyder LB, Mans C. Comparison of dexmedetomidine-ketamine with isoflurane for anesthesia of chinchillas (*Chinchilla lanigera*). *J Am Assoc Lab Anim Sci.* 2016;55:312–316.

4. Parkinson L, Mans C. Anesthetic and postanesthetic effects of alfaxalone-butorphanol compared with dexmedetomidine-ketamine in chinchillas (*Chinchilla lanigera*). *J Am Assoc Lab Anim Sci.* 2017;56:290–295.

5. Doss GA, Mans C, Stepien RL. Echocardiographic effects of dexmedetomidine-ketamine in chinchillas (*Chinchilla lanigera*). *Lab Anim.* 2017;51:89–92.

6. Mans C, Donnelly TM. Update on diseases of chinchillas. *Vet Clin North Am Exot Anim Pract.* 2013;16:383–406.

7. Mans C, Jekl V. Anatomy and disorders of the oral cavity of chinchillas and degus. *Vet Clin North Am Exot Anim Pract.* 2016;19:843–869.

8. Ozawa S, Mans C, Szabo Z, *et al.* Epidemiology of bacterial conjunctivitis in chinchillas (*Chinchilla lanigera*): 49 cases (2005 to 2015). *J Small Anim Pract.* 2017;58:238–245.

9. Martel-Arquette A, Mans C. Management of phimosis and balanoposthitis in a pet chinchilla (*Chinchilla lanigera*). *J Exot Pet Med.* 2016;25:60–64.

10. Martel-Arquette A, Mans C. Urolithiasis in chinchillas: 15 cases (2007 to 2011). *J Small Anim Pract.* 2016;57:260–264.

11. Pignon C, Sanchez-Migallon Guzman D, *et al.* Evaluation of heart murmurs in chinchillas (*Chinchilla lanigera*): 59 cases (1996–2009). *J Am Vet Med Assoc.* 2012;241:1344–1347.

12. Doss G, Mans C, Hoey S, *et al.* Vertebral heart size in chinchillas (*Chinchilla lanigera*) using radiography and computed tomography. *J Small Anim Pract.* 2017;58:714–719.

Speaker Information
(click the speaker's name to view other papers and abstracts submitted by this speaker)

Christoph Mans, DrMedVet, DACZM, DECZM (ZHM)
School of Veterinary Medicine
University of Wisconsin-Madison
Madison, WI, USA

URL: https://www.vin.com/doc/?id=9768913

Respiratory Infections of Exotic Companion Mammals

ExoticsCon Virtual 2020 Proceedings

Dan H. Johnson, DVM, DABVP (Exotic Companion Mammal Practice)
Avian and Exotic Animal Care, Raleigh, NC, USA

Abstract

Respiratory infection is relatively common in pet rabbits, guinea pigs, chinchillas, ferrets, and rats. Predisposing factors include age (i.e., most common in the young and geriatric), stress, improper environment and husbandry, and concurrent illness. Clinical signs include coughing, sneezing, nasal and ocular discharge, and dyspnea. Diagnosis may include radiography, computed tomography (CT), culture/sensitivity, cytology, bacterial polymerase chain reaction (PCR), and necropsy. Common etiologies include various gram-positive and gram-negative bacteria, *Mycoplasma* spp. *Chlamydophila* spp., and viruses; fungal infections are rare. Polymicrobial infections are not unusual. Treatment includes systemic antimicrobials, nebulization, oxygen, supplemental heat, and other supportive care measures. Prognosis varies by etiology and severity of the disease at presentation. Upper respiratory disease carries a better prognosis than pneumonia in most cases.

Introduction

Respiratory infection is a common presenting complaint in exotic companion mammals. It typically occurs in young, sick, debilitated, or immunodeficient animals when natural defense mechanisms have been eroded. Respiratory infection can be precipitated by stress such as shipment, unsanitary conditions, overcrowding, or social conflict among cage mates. Improper environment and husbandry have a significant impact on an individual's resistance to infection. Extremes in temperature, humidity, exposure to waste, and poor nutrition all tend to increase one's susceptibility to it. Likewise, the stress of concurrent infection, advanced age, dental disease, or other concurrent illness can lead to breakdown of the immune system and susceptibility to respiratory infection.

Clinical Signs

Clinical signs of respiratory infection include coughing, sneezing, wheezing, open-mouth breathing, and rapid respiration. Patients with respiratory infection may exhibit hunched posture, unkempt appearance, lethargy, unfocused eyes, disinterest in the surroundings, reduced appetite, weight loss, and/or diarrhea. Increased respiratory effort is usually manifested as pronounced abdominal movement when breathing, and open-mouth breathing occurs in the advanced stages. Discharge from the eyes or nose, and/or diarrhea may be present. A complete physical examination may reveal conjunctivitis, blepharospasm, and ocular or nasal discharge (found either on the face or matted, wet fur on the medial aspect of the forelegs due to facial grooming). Wheezing, crackles, increased bronchovesicular sounds, and/or rales may be ausculted. There may be tachypnea or overt dyspnea, and possibly a fever.

Diagnosis

Diagnosis of respiratory infection in small mammals begins with a thorough history and careful observation of the patient. The clinician should ask about diet, nutritional supplementation, the type of cage, bedding material, how often the cage is cleaned, the presence of other animals, and any new animals that may have been introduced. Also ask about the routine and whether any changes to the routine (e.g., pet sitter, new diet, death of bonded cage mate) have occurred recently.

The workup for upper respiratory infection generally begins with culture and sensitivity from a deep nasal swab but may also include cytology/culture of nasal/nasolacrimal flush, rhinoscopy, skull radiographs, and head CT. Workup for pneumonia generally involves thoracic radiographs. In a normal small mammal, the caudal lung lobes are large and well aerated. Evaluation of the cranial lung lobes may be difficult because in some species (e.g., guinea pigs) they are small. The classic radiographic appearance of bacterial pneumonia is an alveolar pattern, with air bronchograms in severe cases. Lesions can be diffuse, localized to a general area, or lobar in nature. If solitary masses are identified, differential diagnosis should include abscessation or consolidation of lung due to microbial infection.[1]

The hemogram with a respiratory infection may or may not reflect an increase in total white blood cell count, even with pneumonia; a relative neutrophilia and/or lymphopenia are more common. Leucopenia can result from overwhelming bacterial infection, viral infection, or inflammation.

Culture and sensitivity testing should be done before antimicrobial therapy is started. A deep nasal swab is indicated for suspected upper respiratory infection; however, a tracheal wash is generally considered to be more accurate than a nasal swab for diagnosing pneumonia because it can provide material for cytologic exam and bacterial culture.[2] Identification of degenerative neutrophils containing bacterial debris in a tracheal wash specimen is highly supportive of the diagnosis of bacterial pneumonia. While not a widespread practice, transthoracic needle aspiration is another way to obtain lung samples for cytology and culture.[3]

Culture and sensitivity should test for both aerobic and anaerobic organisms, and empirical treatment with broad-spectrum antibiotics should start before results are available. In general, light growths of mixed bacterial populations are less important than the heavy growth of a single species in conjunction with a pathogenic response. Caution must be used when interpreting results; some labs will not report normal flora; however, in an immunocompromised individual, infection may result from normal flora that are opportunistic pathogens. In addition, the normal flora of many species still has not been established or is obscure. If culture cannot be done because antibiotics have already been started, then PCR should be considered. Bacterial PCR can also be helpful in identifying anaerobes (because these frequently do not survive the transport to the lab) and for getting results from an area where bacteria are likely to be dead (e.g., abscess, caseous pus). Where *Chlamydophila* is suspected (i.e., guinea pigs, *C. caviae*), a conjunctival scraping from affected individuals by will contain intracytoplasmic, coccoid, basophilic organisms (i.e., *Chlamydia* elementary and reticulate bodies), or a PCR test is available for this infection, as well.[4]

Necropsy findings from small mammals with respiratory infection will vary by duration and severity of the disease, and by the organism(s) involved. Upper respiratory infections will cause inflammation, pus, and mucous accumulation in the nasal passages, sometimes with focal areas of hemorrhage, abscessation, or necrosis. Mild and acute cases of pneumonia will result in lung congestion, and atelectasis. More severe and chronic pneumonia cases will exhibit suppurative lesions, fibrin adhesions and fibrosis. Severely affected individuals can develop pulmonary abscesses, granulomas, and consolidation. There may also be other organ involvement: lymphadenitis, myocarditis, peritonitis, meningitis, septicemia, tracheitis, bronchitis, otitis media, etc.[5,6]

There are several non-infectious conditions that can cause respiratory distress in exotic companion mammals including nasal obstruction, heat stress, diaphragmatic hernia, pregnancy toxemia, and gastric torsion. Dyspnea and weakness may also be found with heart failure or pulmonary neoplasia.

Treatment

The ideal antibiotic treatment plan for bacterial respiratory infection in small mammals will provide "four-quadrant" coverage, will be bactericidal, easy to administer, safe, and not cause gastrointestinal disease. Antibiotics with a post-antibiotic effect (e.g., aminoglycosides, fluoroquinolones) are preferred, so "pulse therapy" can be employed. Post-antibiotic effect permits once a day dosing, which is less stressful to the patient and improves owner compliance. Aminoglycosides are potentially nephrotoxic; therefore, supplemental fluids are advisable.

As hindgut fermenters, small herbivores rely on active cecal flora for digestion. If antibiotics with a gram-positive spectrum are given, dysbiosis and the overgrowth of pathogenic bacteria will likely occur. Thus, the antibiotics amoxicillin, ampicillin, clindamycin, lincomycin erythromycin are avoided when treating small herbivores. The antibiotics least likely to affect cecal microorganisms include trimethoprim-sulfa, fluoroquinolones, chloramphenicol, aminoglycosides, and metronidazole. Antibiotics that pose an intermediate risk include oral cephalosporins, tetracycline, and doxycycline. In small mammals with a simple digestive tract (e.g., ferret, sugar glider, hedgehog), or minimal cecal fermentation (e.g., rat), the risk of antibiotic-associated gastrointestinal disease is minimal or greatly reduced.

Nebulization therapy is an important method of getting moisture and medication into the nasal passages, trachea, bronchi, and small airways.[7] To adequately treat lower airways, nebulizers should secrete particles between 0.5 and 3.0 micrometers (a room humidifier will not suffice). Saline nebulization alone is helpful. When patients get dehydrated, the mucociliary escalator (ME) of the lower airways becomes impaired. The ME functions to trap particulates and bacteria and moves them craniad by the movement of cilia, to the oropharynx, where they can be coughed up and swallowed. The mucus layer is made up of two layers—the sol, which is watery, where the cilia move, and the gel lying on top, which traps particles. If the sol layer is depleted through dehydration, the cilia become trapped in the gel layer and movement is impeded, inhibiting the escalator. Systemic fluids and airway nebulization can contribute to the effective action of the mucociliary escalator by allowing the sol layer to perform as required. Nebulization can be used to deliver antibiotics directly to the airway surface, bypassing systemic circulation and minimizing side effects. Sometimes other agents are added: mucolytics (N-acetylcysteine), or bronchodilators (aminophylline, terbutaline). F10 Antiseptic Solution (www.F10products.co.uk) is a benzalkonium chloride/hexamethylene biguanide solution with topical antifungal and antibacterial activity. F10 may be added to saline at a rate of 1:250 and is recommended by many for nebulization 2–3 times a day.[8] Use caution when nebulizing products that were not intended for that purpose (e.g., injectable Baytril), because some animals may respond to the ingredients with bronchoconstriction.[9]

There is a blood–bronchus barrier that limits penetration of drugs into the airway secretions much the same way the blood–brain barrier or the blood–prostate barrier prevents drug penetration into those tissues. Penetration into airway secretions is important since many bacterial airway infections remain largely on the luminal airway surface. Nebulization therapy bypasses the "blood–bronchus barrier" and provides topical treatment to the airway. It is important to remember than the major function of the respiratory defenses is the removal of particulates, so nebulized drugs are efficiently removed, and only a small portion of the administered medication reaches the lower airways. Therefore, it is also important to choose a systemic antibiotic that can penetrate the barrier. Lipid-soluble compounds are better able to penetrate the barrier and reach adequate concentrations at the airway surface: metronidazole, chloramphenicol, azithromycin, tetracycline, doxycycline, and fluoroquinolones penetrate the barrier better than penicillin-based antimicrobials; cephalosporins and aminoglycosides have intermediate penetration. Cough suppressants should be avoided when treating pneumonia, since the goal of treatment is to break up and eliminate airway debris and mucus. Gentle coupage (physiotherapy), frequent turning of patients, short walks, and mild to moderate exercise may help to encourage the clearance of sputum.

There are few specific therapies for viral pneumonia. Most viral diseases of rodents do not cause severe disease but can do so in very young animals or those carrying potential secondary bacterial pathogens. Polymicrobial infections are common and, as is the case with pneumonia in most other animals, in small mammals it can involve one or more bacteria or viruses. In many instances where bacteria and virus are both isolated, bacteria are secondary invaders. Provide antibiotics to prevent or treat the secondary bacterial infection, and provide supportive care measures.

Good supportive care is important to treatment success. Exotic companion mammals often present in an advanced state of disease and may not tolerate treatment unless initially stabilized with oxygen, fluids, warmth, and nutritional support. Oxygen supplementation should be provided when there is detectable tachypnea, dyspnea, cyanosis, open-mouth breathing, or when pulse oximeter readings fall below 94%. Oxygen can be provided in an oxygen cage for small animals. It should be humidified prior to use and is toxic over time. The maximum inspired oxygen concentration for long term use is 40%; higher levels may be used for two days or less.

Many exotic companion mammals are prey animals by nature and are nervous in captivity. Many small mammals, especially guinea pigs, are creatures of habit and do not handle changes in routine as well as dogs and cats. Small mammals should be hospitalized in a quiet area away from cats, barking dogs, and other loud noises. Provide a hide box, towel, or other shelter. Animals that are socialized and well adapted to their environment tend to have a much better response to hospital care. Bonded pairs may need to be kept in the hospital together in order to reduce stress. Probiotics have no proven benefits but may help to promote and reestablish healthy gut flora when treating with antibiotics.

Comments on Particular Species and Etiologies

Rabbits

Respiratory disease in rabbits is often caused by bacterial agents: *Pasteurella multocida, Bordetella bronchiseptica, Staphylococcus aureus,* and many others. Reports of viral pneumonia in pet rabbits are rare. Anorexia, weight loss, depression, and rapid fatigue are nonspecific signs, but in a rabbit should raise suspicion of lower respiratory tract disease. Rabbits often appear relatively normal, even with minimally functioning lungs.[6,10]

Pasteurella infection can be chronic or acute. Chronic disease is likely to take the form of upper respiratory disease ("snuffles"). Runny eyes, runny nose, and sneezing can be recurrent, and antibiotics may only provide temporary relief. *Pasteurella* also causes pleuropneumonia or pericarditis, with abscesses developing in or around the lungs.

Bordetella is a common inhabitant of the respiratory tract of rabbits. Its prevalence increases with age. *Bordetella bronchiseptica* infection risk is greatest in young rabbits and in hosts with compromised immune function. *Bordetella* has local-acting cytotoxic effects that impair host defenses and may be an important predisposing factor in *Pasteurella* infections.

Staphylococcus aureus can be isolated from the respiratory tract of both healthy and diseased rabbits. While it is probably a secondary invader of compromised mucosa, *S. aureus* produces toxins that can block several host defenses. Like *Pasteurella,* disseminated infection can cause pneumonia and abscessation of the lungs and heart.

Guinea Pigs

Bacterial respiratory infection is considered to be one of the most significant diseases affecting guinea pigs. Pneumonia is the number one cause of death among guinea pigs in some surveys. Young guinea pigs are particularly at risk for infection. Many of the organisms that cause respiratory disease are acquired by guinea pigs as babies while they are still in the breeding colony, protected by maternal antibodies. Those with subclinical infection often continue to carry these pathogens, and only develop clinical disease later, when stress or concurrent illness occurs.

The course of infection can vary from mild upper respiratory infection with subtle clinical signs, to rapidly progressive and fatal pneumonia, associated with respiratory failure and/or sepsis. Guinea pigs are by nature very stoic and this combined with their natural tendency to hide disease means pneumonia is often advanced by the time clinical signs are noticed.

Rhinitis and pneumonia are usually associated with opportunistic microbes. The two most important pathogens are the bacteria *Bordetella bronchiseptica* and *Streptococcus pneumoniae.* Other common bacteria include *Klebsiella pneumonia, Streptobacillus moniliformis, Staphylococcus aureus, E. coli, Pasteurella pneumotropica, Pasteurella multocida, Streptococcus zooepidemicus, Streptococcus pyogenes, Citrobacter freundii, Yersinia pseudotuberculosis,* and *Pseudomonas aeruginosa. Chlamydophila caviae* usually causes mild, self-limiting conjunctivitis but can also contribute to stress and lower respiratory tract bacterial infections.[6,11]

Guinea pigs may also develop viral pneumonia. Guinea pig adenovirus (GPAdV) can cause severe, necrotizing bronchopneumonia with high mortality. Parainfluenzavirus has also been shown to cause disease. Pneumonia in guinea pigs also can be caused by atypical organisms including *Pneumocystis carinii* and nematode parasites. Vitamin C should be supplemented, initially by injection 100 mg/kg subcutaneously, and then provided daily through hand feeding and fresh vitamin C-rich vegetables and fruit. Guinea pigs can be vaccinated against *Bordetella* with commercially available canine or porcine vaccines.[5,6]

Chinchillas

Respiratory infections in chinchillas are relatively uncommon. Affected individuals usually are immunocompromised by age, nutritional status, or husbandry-related stress (overcrowding, high humidity, poor ventilation). Pneumonia in the chinchilla tends to be chronic, resulting in ocular or nasal discharge, lymphadenopathy, dyspnea, anorexia, depression, poor hair coat, and weight loss. Likely pathogens include *Pasteurella, Bordetella, Streptococcus, Klebsiella,* and *Pseudomonas.* Viral respiratory disease of chinchillas is not reported. Antibiotic combinations may be indicated. Prevention is by correcting husbandry and reducing stress.[11]

Ferrets

Respiratory infections are uncommon. Viral causes include canine distemper and influenza virus. Bacteria reported to cause primary infection include *Streptococcus zooepidemicus, S. pneumoniae,* and groups C and G streptococci. Bacterial rhinitis and conjunctivitis, primarily seen in young, newly acquired animals usually responds to treatment with amoxicillin. Bacterial pneumonia in ferrets is typically suppurative, affecting the bronchial airways, the lung lobes, or both. Gram-negative bacterial pneumonia has been documented with *Escherichia coli, Klebsiella pneumoniae, Pseudomonas aeruginosa, Bordetella bronchiseptica* and *Listeria monocytogenes.* Fungal infections (*Pneumocystis carinii,* blastomycosis and coccidiomycosis) are uncommon in pet ferrets. A novel *Mycoplasma* sp. causes a chronic bronchitis and persistent cough, with eventual death due to lung pathology.[12,13]

Rats

Respiratory disease caused by infectious agents is the most common health problem in rats. Three major respiratory pathogens cause overt clinical disease: *Mycoplasma pulmonis*, *Streptococcus pneumoniae*, and *Corynebacterium kutscheri*. Two additional bacteria (cilia-associated respiratory [CAR] bacillus and *Haemophilus* spp.) and three viruses (Sendai virus [a parainfluenza virus], pneumonia virus of mice [a paramyxovirus], and rat respiratory virus [a hantavirus]) are minor respiratory pathogens that by themselves rarely cause overt disease.[6,14] A coronavirus of rats, sialodacryoadenitis (SDA) virus, is highly infectious and causes overt disease confined to the eyes, ears, nose and throat. These minor respiratory pathogens interact synergistically as co-pathogens with the major respiratory pathogens to produce two major clinical syndromes: chronic respiratory disease (CRD) and bacterial pneumonia.

CRD is also known as murine respiratory mycoplasmosis because *M. pulmonis* is the major component of the disease.[15] Clinical signs are highly variable, and initial infection occurs without any clinical signs. Both upper and lower respiratory tracts are involved. Signs may include snuffling, nasal discharge, red tears, rapid respiration, weight loss, hunched posture, ruffled coat, and head tilt. CRD varies greatly in disease expression because of many environmental, host, and organismal factors that influence the host–pathogen relationship: cage ammonia levels, concurrent infection (Sendai virus, SDA virus, pneumonia virus of mice, rat respiratory virus, CAR bacillus), genetic susceptibility of host, virulence of *Mycoplasma* strain, and vitamin A or E deficiency. Antibiotic therapy (enrofloxacin/doxycycline combination, azithromycin) will not cure CRD but may alleviate clinical signs; affected animals typically have persistent *M. pulmonis* infection. Removing cage litter and replacing with clean paper daily may help reduce ammonia levels. Bronchodilators and short-acting corticosteroids are also sometimes helpful.

Bacterial pneumonia, the other major clinical syndrome, is nearly always caused by *Streptococcus pneumoniae,* usually with the help of *M. pulmonis,* Sendai virus, or CAR bacillus coinfection. *Corynebacterium kutscheri* causes pneumonia only in severely immunosuppressed individuals and is rare in pet rats. Pneumonia caused by *S. pneumoniae* can be of sudden onset. Young rats are more severely affected than older ones, and the only symptom may be sudden death. Signs in mature rats include dyspnea, snuffling, abdominal breathing, and purulent nasal exudate (nares and front paws). Tentative diagnosis can be made by a Gram stain of this exudate (will reveal numerous gram-positive diplococci). Because severe bacteremia and multiorgan abscesses/infarctions are common; antibiotic treatment must be aggressive. Beta-lactamase-resistant penicillins are recommended.

Prognosis

The prognosis for respiratory infection varies by etiology and severity of the disease at presentation. Most small herbivores are obligate nasal breathers; thus, blockage of the nasal passages can be deadly. Severely affected animals resort to open-mouth breathing, which reduces food and water intake and worsens overall condition. For this reason, even upper respiratory infections are a serious matter in most exotic companion mammals; however, upper respiratory infection can usually be managed with appropriate care. Pneumonia, on the other hand, is often difficult to reverse. Affected animals often have underlying immunodeficiency and tend to decline rather than recover. Prognosis becomes guarded once there is obvious respiratory distress.

Prevention

Many of the causes of respiratory infection in small exotic mammals are husbandry related, and corrections need to be made in order to prevent disease. Keeping a closed colony will ensure that new diseases are not introduced. New arrivals should go into quarantine, and they should not be mixed with the general population until there has been a reasonable quarantine period, usually 30–90 days.

References

1. Capello V, Lennox AM. Diagnostic imaging of the respiratory system in exotic companion mammals. *Vet Clin North Am Exot Anim Pract.* 2011;14:369–389.

2. Mancinelli E. Respiratory disease in rabbits. *In Pract.* 2019;41:121–129.

3. Chockalingam A, Hong K. Transthoracic needle aspiration: the past, present and future. *J Thorac Dis.* 2015;7(4):292–299.

4. Colby LA, Nowland MH, Kennedy LH. *Clinical Laboratory Animal Medicine: An Introduction.* 5th ed. Ames, IA: Wiley-Blackwell; 2020:212–242.

5. Oglesbee BL. *Blackwell's 5-Minute Veterinary Consult: Small Mammal.* Ames, IA: Wiley-Blackwell; 2011:171–172, 309–311, 476–478, 597–599.

6. Quesenberry KE, Carpenter JW, eds. *Ferrets, Rabbits and Rodents: Clinical Medicine and Surgery.* 3rd ed. St. Louis, MO: Elsevier Saunders; 2012:78–85, 205–216, 298–299, 315, 362–363.

7. Powers LV. Techniques for drug delivery in small mammals. *J Exot Pet Med.* 2006;15(3):201–209.

8. Varga M, ed. *Textbook of Rabbit Medicine.* St. Louis, MO: Elsevier Saunders; 2014:171, 395–397.

9. Keeble E, Meredeth A, eds. *BSAVA Manual of Rodents and Ferrets.* Gloucester, England: British Small Animal Veterinary Association; 2009:142–149, 288–290.

10. Johnson-Delaney CA, Orosz SE. Rabbit respiratory system: Clinical anatomy, physiology, and disease. *Vet Clin North Am Exot Anim Pract.* 2011;14:257–266.

11. Yarto-Jaramillo E. Respiratory system anatomy, physiology, and disease: guinea pigs and chinchillas. *Vet Clin North Am Exot Anim Pract.* 2011;14:339–355.

12. Kiupel M, Desjardins, DR, Lim A, *et al.* Mycoplasmosis in ferrets. *Emerg Infect Dis.* 2012;18:1763–1770.

13. Richardson J, Perpinan D. Disorders of the respiratory tract. In: Johnson-Delaney CA, ed. *Ferret Medicine and Surgery.* Boca Raton, FL: CRC Press; 2017:311–324.

14. Kling MA. A review of respiratory system anatomy, physiology, and disease in the mouse, rat, hamster, and gerbil. *Vet Clin North Am Exot Anim Pract.* 2011;14:287–337.

15. Graham JE, Schoeb TR. Mycoplasma pulmonis in rats. *J Exot Pet Med.* 2011;20(4):270–276.

Speaker Information

(click the speaker's name to view other papers and abstracts submitted by this speaker)

Dan H. Johnson, DVM, DABVP (Exotic Companion Mammal Practice)
Avian and Exotic Animal Care
Raleigh, NC, USA

URL: https://www.vin.com/doc/?id=9768917

Antimicrobial Resistance and Therapy in Exotic Pet Practice

ExoticsCon Virtual 2020 Proceedings

Andrew Bean, DVM, MPH, DABVP (Exotic Companion Mammal Practice)
Avian & Exotic Medicine Service, Animal Emergency & Referral Center of Minnesota, Oakdale, MN, USA

Abstract

Antimicrobial resistance is a major public health crisis. Antimicrobial stewardship programs and guidelines for judicious use of antimicrobials have become commonplace in human and small animal medicine, but the practice of exotic animal medicine lacks such guidance. This master class will review the current situation of the antimicrobial resistance crisis, mechanisms for the spread of resistance, nuances of diagnosing and treating bacterial infections, and review general guidance for antimicrobial stewardship.

Introduction

Antimicrobial resistance is a major public health crisis. Bacteria, fungi, and parasites are all rapidly acquiring the ability to defeat the drugs designed to destroy them. In human health, **at least** an estimated 2,868,700 people are sickened by an antimicrobial-resistant infection ... in the United States ... each year. An estimated 35,900 of these people will die. *Clostidioides difficile* (formerly *Clostridium difficile*) infections related to antimicrobial use and resistance account for an additional 223,900 cases and 12,800 deaths annually.[1] Put those numbers together and we find that over 3 million people get infections related to overuse of antimicrobials, and nearly 50,000 of them will die in the United States each year.

The days of the miraculous cure with the first dose of antibiotics are gone. "Stop referring to a coming post-antibiotic era," writes Robert Redfield, director of the United States Centers for Disease Control & Prevention, "it's already here. You and I are living in a time when some miracle drugs no longer perform miracles and families are being ripped apart by a microscopic enemy. The time for action is now, and we can be part of the solution."[1]

Antimicrobial resistance is an ecological problem. With microbes inhabiting almost all corners of our world and our bodies, exposure to antimicrobials does not stop outside of a living animal. Drugs may leak into the environment at the manufacturing plant.[2] Antimicrobials administered to patients are eventually excreted in urine or feces, leading to direct environmental exposure or concentration in wastewater treatment facilities. Surface water run-off and effluent from water treatment plants allows the drugs to concentrate in natural bodies of water.[3] Reservoirs of antimicrobial resistance genes have been found in environmental bacteria[4,5] and in organisms used as biopesticides[6].

Environmental microbes may be exposed to antimicrobials, causing a selection pressure to express a variety of antimicrobial resistance genes. While microbes in the environment often naturally produce antimicrobial substances, these do not appear to have a substantial effect on overall expression of resistance genes. It is the additional selection pressure exerted by manufactured antimicrobials that results in higher resistance gene expression.[7] Antimicrobials, bacteria, and resistance genes move between the compartments of animals (including humans), soil, and water, creating conditions in which resistant bacteria may emerge and proliferate.[8]

There are two ways that microbes acquire resistance genes: mutation and horizontal transfer. Horizontal transfer of resistance genes overwhelmingly occurs via plasmid, a "raft" of genetic material that encodes for a specific trait.[9] This allows the spread of antimicrobial resistance genes across multiple species of bacteria. The sharing of resistance genes in the environment may flow both ways between human pathogens and non-pathogenic environmental bacteria.[7]

Evidence for the role of point sources (e.g., pharmaceutical plants, wastewater treatment facilities) in development and propagation of antimicrobial resistance in the environment is incomplete. A 2017 systematic review of the topic found that many studies had unclear assessment of bias, limited statistical analysis, and no measurement of effect (e.g., odds ratio, mean difference). The authors noted the difficulty of studying antimicrobial resistance in the environment due to the complex processes, field work, and numerous factors influencing relationships, and provided a framework for further studies of the issue.[10]

It is undeniable that overuse of antibiotics plays a major role in the increasing prevalence of antimicrobial resistance. Within a given species, antimicrobial use is considered to be the main driver of antimicrobial resistance.[11] The influence of these prescriptions on antimicrobial resistance in other species is less well studied, but some examples stand out.

The use of ceftiofur in chicken hatcheries was linked to increased resistance to ceftriaxone in *Salmonella* Heidelberg infection in humans (especially those associated with chicken meat). A temporary ban on using ceftiofur in hatcheries resulted in a reduction of resistance, and reintroduction of ceftiofur use resulted in increased resistance.[12] Avoparcin, a glycopeptide antibiotic previously used as a growth promoter in poultry and swine, has been shown to cause high levels of vancomycin-resistant *Enterococcus faecium* in the feces of treated animals. The resistance was concurrently seen in enterococcal infections in human medicine, where vancomycin is used in treatment of multidrug resistant infections.[13]

Bacteria-Host-Environment Relationship

It is essential to consider the symbiotic relationship between bacteria, the colonized host organism, and the environment as a complex ecosystem. Prokaryotic cells outnumber eukaryotic cells in most animals. Bacteria are essential players in many life functions, including digestion and immune function. To consider bacteria dichotomously as either "good" or "bad" is to vastly oversimply the relationship.[14]

Bacteria, like all forms of life, seek to survive and reproduce. In general, host organisms allow bacteria to do this and variably extract some benefit from the relationship. In some cases, the organism finds itself in a niche that causes disease to the host. Often the host environment plays a role in allowing this to occur. This scenario is often seen in exotic pets that have substandard husbandry; the host ecosystem is disrupted, and disease ensues. Unless the underlying problems are corrected, disease may recur after initially successful treatment.

Many factors that can influence the development of microbial disease in a patient. A partial list includes:

- Housing density
- Environmental temperature
- Environmental humidity
- Environmental lighting
- Air quality
- Nutrition
- Opportunities to engage in natural behaviors
- Housing substrate
- Perception of safety

The administration of antimicrobials may have substantial impact on a patient's microbial ecosystem. Most systemic antimicrobials diffuse throughout the body in varying concentrations, exposing far more than the targeted microbes. The consequences of this may be rapid and life threatening, as seen in antibiotic-associated dysbiosis in hindgut fermenters or *C. difficile* infections in humans. However, more long-term effects may be seen: recent administration of antimicrobials in dogs increases the likelihood of carrying methicillin-resistant *Staphylococcus pseudintermedius*;[15] multidrug resistance may be induced in *E. coli* present in the gastrointestinal tract of dogs[16,17]. In addition, idiosyncratic drug reactions, hypersensitivity reactions, accidental overdose, stress from administration, and pain from injections may occur and have a negative impact on health status.

Diagnostic Considerations

Ockham's razor— "The simplest explanation is the most likely explanation"—becomes a trap in exotic animal medicine. The sneezing rabbit must be infected with *Pasteurella multocida*.

The alopecic guinea pig has follicular ovarian cysts. The dyspneic bird has bacterial pneumonia or air sacculitis. While these diagnoses may be commonly reported in the overall patient population, it is important to remember Hickam's dictum when considering the individual animal: "[The patient] may have as many diseases as [it] damn well pleases."

Every effort must be made to localize a disease and determine its nature and underlying causes. In exotic medicine, a thorough review of the husbandry is essential. Do not just ask, "What?" but also, "How?" and "When?" For example:

- What is the temperature on the cool side of the cage? How do you know? What type of thermometer is used? When was it purchased?
- What is the cage cleaning frequency? How is it performed? What cleaners are used and how are they applied?

Diagnostic testing should be applied in a logical manner. In general, CBC, serum chemistry, and radiography are considered baseline diagnostics and should be recommended in most cases to evaluate for evidence of infection versus other diseases prior to the commencement of therapy. Computed tomography (CT) has become increasingly useful as a first-line imaging technique, providing superior diagnostic information compared to traditional radiography, especially in small patients.

Cytology may be used to guide therapy pending results of culture or PCR. A simple Gram stain and Romanowsky stain (e.g., Diff-Quik) yield basic information that can inform drug selection and provide context for interpretation of culture/PCR results. In general, cytology is considered supportive of bacterial infection if bacteria are visualized within phagocytes (e.g., neutrophils, heterophils). Extracellular bacteria may be pathogens, contaminants, or commensals.

In confronting infectious diseases, PCR and microbial culture become especially important tools for guiding treatment. One must remember that the result is only as good as the sample. Submitting a sample from a contaminated surface may produce a spurious result. Inappropriate handling of a sample (e.g., temperature, transport medium) may lead to a false negative result.

In general, swabs may not be the best tools for acquiring culture specimens, as few of the colony forming units (CFUs) within a swab will actually be cultured. Air trapped between the cotton fibers may inhibit anaerobic growth. Necrotic and purulent debris are similarly suboptimal candidate samples for culture, as they may not contain viable organisms. A piece of infected tissue is generally the preferred sample for culture.[18] For example, abscess capsules (or entire abscesses) are considered to be superior to pus in culturing certain organisms.[19] Tissue culture samples may be submitted in a no-additive tube (e.g., white-top) with one or two drops of sterile saline added to prevent desiccation (if needed). If a tissue sample is submitted in a tube containing transport medium and a swab, it should be clearly indicated on the submission form that the sample being submitted is the tissue, not the swab. Not indicating the thing to be cultured runs the risk of the laboratory disregarding the presence of tissue and simply rubbing the accompanying swab on culture media.

Interpretation of culture results requires nuance and knowledge of the possible normal microbial flora at the site cultured. As mentioned above, cytology can provide an especially useful perspective. Isolation of multiple organisms from potentially contaminated sites may simply represent colonization rather than infection. Pure growth of a single organism is more consistent with the need for antimicrobial therapy. The degree of growth should also be considered, as abundant growth is generally considered more supportive of infection. However, not all bacteria exhibit the same rates of growth, and more rapidly growing bacteria may outcompete slowly growing bacteria for resources. This effect may be reduced by refrigeration of the sample after collection and prompt submission to the laboratory.[18]

Susceptibility results may be determined using methods including disk diffusion, broth dilution, or Epsilon test (E-test), ideally according to standards set by the Clinical and Laboratory Standards Institute (CLSI). Broth dilution and E-testing are generally preferred, as they provide a quantitative concentration of drug that will inhibit the isolate. This can then be used to guide drug dosing (see Treatment Considerations). Isolates reported as having a minimum inhibitory concentration (MIC) of less than (<) the lowest tested concentration are generally considered highly susceptible, as growth of the isolate is inhibited by even the lowest concentration of drug tested. Isolates that have an MIC above the lowest tested concentration but less than the MIC breakpoint of a resistant organism have likely undergone some of the initial steps towards developing resistance, and continued progress toward full resistance may be more likely for these isolates.[18]

Treatment Considerations

"Many veterinarians have the mentality that in exotics species ... enrofloxacin is a panacea. The routine use of advanced, broad spectrum antibiotics implies a low level of skill on the part of the clinician. It is our duty to develop appropriate antimicrobial stewardship. Veterinarians should limit antibiotic use to those cases in which their use is warranted."—Sean Perry and Mark Mitchell[20]

Treatment with antimicrobials should be reserved only for cases in which there is evidence—not just suspicion—of a bacterial infection. Viral infection, immune-mediated disease, and non-infectious inflammation may all appear grossly similar to bacterial infection. The drug chosen should ideally be based on PCR or culture/sensitivity results. If therapy must be initiated prior to results of these tests, choose a relatively narrow-spectrum drug that will affect the suspected pathogen(s) and a route of administration that will have the least impact on the body's microflora.[21]

Categorization of antimicrobials into tiers to guide selection have been proposed and designed, often around specific diseases.[22,23,24] Efforts to produce similar guidelines for exotic species have been few. The Zoological Medicine Service at the University of Georgia devised antimicrobial tiers for exotic patients based on retrospective evaluation of their cases (Table 1).[25] This should be considered but not as absolute, as infection agent prevalence and susceptibility patterns may differ dramatically based on region and caseload.

Table 1. University of Georgia Zoological Medicine Service antibiotic tiers[25]

Tier 1	
Uses	**Drugs**
1. Prevention of infection in severely compromised animals 2. Intraoperative IV or IO administration to prevent infection (e.g., orthopedics) 3. First line therapeutic antimicrobial, pending culture and sensitivity results	Potentiated sulfonamides (e.g., sulfamethoxazole/trimethoprim) Tetracyclines Basic/potentiated penicillins (e.g., penicillin, ampicillin, amoxi/clav) Metronidazole Lincosamides (lincomycin, clindamycin) Aminoglycosides First and second generation cephalosporins First generation quinolones (e.g., oxolinic acid, nalidixic acid)
Tier 2	
Uses	**Drugs**

Only to be used if sensitivity/MIC indicates Tier 1 options are ineffective	Third generation cephalosporins (e.g., ceftazidime, ceftiofur) Advanced penicillins (e.g., piperacillin, carbenicillin, ticarcillin) Second generation fluoroquinolones (e.g., enrofloxacin, marbofloxacin) Florphenicol, chloramphenicol
Tier 3	
Uses	**Drugs**
Only to be used in cases of multidrug resistance where specified hospital criteria have been met and authorization has been granted for their use	Glycopeptides (e.g., vancomycin) Carbapenems (e.g., imipenem) Oxazolidinones (e.g., linezolid) Fourth generation & above cephalosporins (e.g., cefepime) Ketolides (e.g., telithromycin) Lipopeptides (e.g., daptomycin) Ansamycins (e.g., geldanamycin) Third generation & above fluoroquinolones (e.g., levofloxacin)

Drug selection should also consider the importance of antimicrobials to human medicine. The World Health Organization (WHO) maintains a list of drugs considered critical for human medicine, divided into the three major tiers—Important, Highly Important, and Critically Important—based on five criteria. The Critically Important category is further subdivided into High Priority and Highest Priority subdivisions (Table 2).[26]

Table 2. WHO critically important antimicrobials for human medicine[26]

Category	Drug classes
Important	Aminocyclitols Cyclic polypeptides Nitrofurantoins Nitroimidazoles (e.g., metronidazole) Pleuromutilins
Highly important	Amidinopenicillins Amphenicols First and second generation cephalosporins and cephamycins Penicillins (anti-staphylococcal) Pseudomonic acids Riminofenzaines Steroid antibacterials Streptogramins Sulfmonamides, dihydrofolate reductase inhibitors, and combinations thereof Sulfones Tetracyclines
Critically important—High priority	Aminoglycosides Ansamycins Carbapenems and other penems Glycylcylines Lipopeptides Monobactams Oxazolidinones Penicillins (natural, aminopenicillins, antipseudomonal) Phosphoric acid derivatives Drugs used solely to treat tuberculosis or other mycobacterial diseases
Critically important—Highest priority	Cephalosporins (Third generation and above) Glycopeptides Macrolides and ketolides Polymixins Quinolones

The World Organization for Animal Health (OIE—Office International des Epizooties) has also categorized antimicrobials into three tiers: Veterinary Important, Veterinary Highly Important, and Veterinary Critically Important (Table 3). These categorizations were made with an emphasis on production animal medicine; companion animals do not appear to have been a major consideration.[27]

Table 3. OIE categorization of veterinary important antimicrobials for food-producing animals[27]

Category	Drug classes
Veterinary important	Aminocoumarin (e.g., novobiocin) Arsenicals (e.g., nitarsone) Bicyclomycins (e.g., bicozamycin) Fusidanes (e.g., fusidic acid) Orthosomycins (e.g., avilamycin) Quinoxalines (e.g., carbadox) Streptogramins (e.g., virginiamycin) Thiostreptons e.g., nosiheptide)
Veterinary highly important	Ansamcin—Rifamycins (e.g., rifampicin) First and second generation cephalosporins Ionophores Lincosamides Phosphoric acid derivatives (e.g., Fosfomycin) Pleuromutilins (e.g., tiamulin) Polypeptides (e.g., bacitracin, gramicidin) Polymixins First generation quinolones (e.g., nalidixic acid)
Veterinary critically important	Aminocyclitol Aminoglycosides Amphenicols Third and fourth generation cephalosporins Macrolides Penicillins (all) Second generation quinolones (e.g., enrofloxacin, marbofloxacin) Sulfonamides (alone and potentiated) Tetracyclines

The OIE also notes that fluoroquinolones and third and fourth generation cephalosporins, along with colistin, are considered critically important for both human and animal health. Therefore, the OIE has issued the following guidelines for use of these drugs (underling added by the author; two additional guidelines are omitted as the author deems them not applicable to exotic animal practice):[27]

- Not to be used as preventive treatment applied by feed or water in the absence of clinical signs in the animals to be treated
- **Not to be used as a first line treatment unless justified**; when used as a second line treatment, it should ideally be based on the results of bacteriological tests.

Some of the key take-home points from the above tables and lists:

- Enrofloxacin is not a first line drug
- Third generation cephalosporins (e.g., ceftazidime) are not first line drugs
- Enrofloxacin is not a first line drug
- Some acceptable first-line drugs include trimethoprim/sulfa combinations, doxycycline, first generation cephalosporins, and basic potentiated penicillins (ampicillin, amoxicillin/clavulanate)
- Enrofloxacin is not a first line drug
- Third generation cephalosporins (e.g., ceftazidime) are not first line drugs
- Enrofloxacin is not a first line drug

Whenever possible, drug dosing should be guided by published pharmacokinetic and pharmacodynamic studies rather than anecdotal use and perceived success. Without pharmacokinetic guidance, the practitioner has no real idea if the chosen therapy will even reach therapeutic concentrations at the site of infection. The most up-to-date editions of common exotic animal formularies should be used, and the cited source of prospective dosing regimens assessed.

Antimicrobials are generally grouped into two classifications based on their effect on bacteria: those that simply inhibit the growth of microbes (i.e., bacteriostatic), and those that actively kill microbes (i.e., bactericidal) (Table 4). This stratification is based on *in vitro* models and may not reflect the way the drug performs in the patient at the site of infection. Certain bacteriostatic drugs (e.g., macrolides) may reach higher concentrations in white blood cells and at sites of infection, resulting in a bactericidal effect.

Table 4. Antimicrobial drug classes separated into bacteriostatic and bactericidal groups[18]

Bacteriostatic	Bactericidal
Tetracyclines Phenicols Macrolides* Lincosamides* Unpotentiated sulfonamides	Penicillins Cephalosporins Fluoroquinolones Metronidazole Aminoglycosides Potentiated sulfonamides Polymixin Colistin

	Rifampin Glycopeptides

* = accumulation in white blood cells may allow achievement of bactericidal concentrations at site of infection.

Antimicrobials may also be classified based on how their effect relates to the concentration of drug (Table 5). Antimicrobials whose efficacy is best determined by their maximum plasma concentration compared to the MIC are considered concentration dependent. Drugs whose effect is best determined by the amount of time plasma drug concentrations remain above the MIC are considered time dependent.

Table 5. Antimicrobial drug classes separated into concentration dependent and time dependent groups[18]

Concentration dependent	Time dependent
Aminoglycosides Fluoroquinolones Metronidazole Azithromycin Ketolides	Beta-lactams Glycopeptides Macrolides Tetracycline Lincosamides

The goal of antimicrobial therapy is to achieve sufficient concentrations of the drug at the site of infection in order to kill the infecting organisms, while concurrently avoiding the patient side-effects of the drug.[18] Culture and susceptibility results providing an MIC for each isolate may be used in the formulation of an antimicrobial therapeutic regimen, especially by comparing the MIC to the expected maximum plasma concentrations of the drug (C_{max}) or the area under the concentration-time curve (AUC_{24}). Guidelines exist for dosing some antimicrobial agents:[18,28]

- Aminoglycosides: the C_{max}-to-MIC ratio should be between 8 and 10 (C_{max}:MIC = 8 to 10)
- Fluoroquinolones:
  - The ratio of the C_{max} to the MIC should be between 8 and 10 (C_{max}:MIC = 8 to 10)

    or
  - The ratio of the area under the concentration-time curve to the MIC should be 100 to 125 (AUC_{24}:MIC = 100 to 125)
- Beta-lactams (penicillins, cephalosporins, monobactams, carbapenems)
  - The ratio of the C_{max} and the MIC should be 2 to 4

    and
  - The plasma drug concentration should remain above the MIC for as long as possible, ideally at least 50% of the dosing interval

The route of administration may substantially influence the chances for development of resistance. Topical medication, whether applied directly in the form of creams and ointments, or inhaled via nebulization, allows specific delivery of mediation to the intended area with minimal impact on the host microbiome. Care should be used in selecting topical treatment of an area that the patient can lick, and medications that will not cause harm if ingested should be used.

The oral route is commonly employed but is not always the most appropriate. Interspecies differences in gastrointestinal anatomy and physiology result in highly variable bioavailability.[28] Orally administered antibiotics result in direct exposure of the gastrointestinal microflora to the chosen drug, which may cause side effects ranging from mild gastroenteritis to death (especially in hindgut fermenter species—rabbits, guinea pigs, chinchillas, hamsters, etc.).[29] Furthermore, the patient spitting out a portion of the drug may result in subtherapeutic dosing. Reptile patients being medicated by mouth should be maintained within their preferred optimal temperature range to ensure appropriate gastrointestinal function and metabolism.[30]

The subcutaneous and intramuscular routes, while less commonly used, can circumvent the first-pass metabolic effect and deliver drug into the circulation. The volume to be administered is often small, and owners may be instructed to perform administration at home. Care should be taken not to administer injectable medications into the caudal half of the body of birds and reptiles as the renal and hepatic portal systems may shunt drug through the kidneys or liver prior to systemic distribution. While the significance of this shunting varies widely by the drug, in general the practice is advised to avoid inadvertent weakening of therapy.[31]

Antimicrobial Stewardship

Antimicrobial stewardship practices have gained popularity in human and small animal medicine but are lacking in exotic animal medicine. The American Veterinary Medical Association (AVMA) has defined antimicrobial stewardship as the actions veterinarians take individually and as a profession to preserve the effectiveness and availability of antimicrobial drugs through conscientious oversight and responsible medical decision-making while safeguarding animal, public, and environmental health.

A number of authors have outlined the core principles of antimicrobial stewardship in veterinary medicine. One group describes the principles as the 5 Rs:[32]

- Responsibility
- Reduction
- Replacement
- Refinement
- Review

The AVMA has developed a similar but more detailed list of core principles and the components of each principle (Table 6). A checklist based on these principles is available to aid practitioners in developing and implementing antimicrobial stewardship policies in their practices.[a]

Table 6. AVMA's core principles of antimicrobial stewardship in veterinary medicine[33]

Core principle	Components
Commitment to stewardship	Engage all practice members and relevant stakeholders in the stewardship effort. Develop stewardship plans that incorporate dedication to and accountability for disease prevention and that also optimize the prescribing, administration, and oversight of antimicrobial drugs. Identify high-priority conditions that are commonly treated with antimicrobial drugs on which to focus stewardship efforts.

	Demonstrate commitment to systematically assessing the outcomes of antimicrobial drug therapy. Identify one or more individuals to lead the stewardship plan and provide accountability.
Advocate for a system of care to prevent common diseases	Identify barriers to improving disease prevention. Work with clients to adopt preventive and management strategies to minimize the need for antimicrobial drugs. These strategies include animal husbandry and hygiene, biosecurity and infection control, nutrition, and vaccination programs. Consider alternatives to antimicrobial drugs.
Select and use antimicrobial drugs judiciously	Identify barriers to appropriate antimicrobial prescribing and usage. Use an evidence-based approach for making a diagnosis and determining whether an antimicrobial drug is indicated. Make an informed selection of an appropriate antimicrobial drug and regimen. Refer to relevant veterinary medical guidelines for judicious therapeutic use. Assess outcomes of antimicrobial use.
Evaluate antimicrobial drug use practices	Encourage development of a program for the evaluation of antimicrobial drug prescribing at the veterinary practice or aggregated levels. Ensure that feedback is provided to veterinarians. Support analyzing and sharing of antimicrobial drug use data while preserving veterinarian-client confidentiality. Engage clients to identify barriers to implementation of stewardship programs and to evaluate antimicrobial storage, administration, and other use practices.
Educate and build expertise	Make resources available and encourage the development of expertise in antimicrobial stewardship. Keep up-to-date on strategies for disease prevention, use of antimicrobial alternatives, and selecting and using antimicrobial drugs. Critically appraise and then implement appropriate existing clinical guidelines for antimicrobial use. Provide client education on antimicrobial stewardship, including conditions when antimicrobial drugs are not needed. Support research on antimicrobial drug use and resistance.

The AVMA has also provided a list of principles for judicious use of antimicrobials (bold performed by the author for added emphasis):[34]

- Antimicrobials should be used in animals only after careful review and consideration of the following points:
  - A general or preliminary diagnosis has been made which indicates that antimicrobial therapy is appropriate.
  - **Culture and antimicrobial susceptibility testing should be performed when possible to guide the selection of antimicrobials**.
  - Regimens for antimicrobial prevention, control and treatment should be established using accepted scientific and clinical principles, such as microbiological and pharmacological tenets.
  - Antimicrobial therapy for uncomplicated viral infections and non-septic inflammatory conditions should be avoided.
  - Duration of therapy should be based on scientific and clinical evidence in order to obtain the desired health outcome while minimizing selection for antimicrobial resistance, as allowed by law.
  - Antimicrobial therapy should be targeted to ill or at-risk animals when possible, and other management strategies should be used to prevent infection in healthy individuals.
- Accurate records of therapy and outcome should be maintained.
- **Environmental contamination with antimicrobials must be avoided whenever possible**.

A variety of interventions to improve antimicrobial prescribing practices have been undertaken in human medicine. A systematic review and meta-analysis of published intervention strategies found an overall 15% improvement in prescribing practices regardless of intervention type. A summary of interventions is provided (Table 7). Interventions were associated with a reduction in length of hospitalization by 1.12 days, and no increase in patient morbidity or mortality. However, one restrictive intervention—requiring prior approval for antibiotic use—showed unintended detrimental consequences: negative professional culture due to breakdown in trust, and delay in time to first antibiotic use (when such use was indicated). Further comparison of studies suggested that interventions featuring some element of enablement provided a greater effect, though this was only supported by some studies. Additionally, interventions that included feedback appeared to be more effective.[35]

Table 7. Definition of behavior change techniques and intervention functions[35]

Intervention function	Definition	Intervention components
Education	Increasing knowledge or understanding	Educational meetings Dissemination of educational materials Educational outreach
Persuasion	Using communication to induce positive or negative feelings or to stimulate action	Educational outreach through academic detailing or review of individual patients with recommendations for change
Restriction	Using rules to reduce opportunity to engage in the target behavior (or increase the target behavior by reducing the opportunity to engage in competing behaviors)	Controlled access to formulary Authorization requirement to use certain drugs Automatic stop orders Therapeutic substitution
Environmental restructuring	Changing the physical context	Physical reminders (e.g., posters, wallet-size summaries, reiteration on lab reports)

Enablement	Increasing means/reducing barriers to increase capability/opportunity	Audit & feedback Decision support through computerized systems or through circumstantial reminders that were triggered by actions or events related to the targeted behavior Educational outreach by review and recommend change

Veterinarians in Australia were recently surveyed regarding barriers to and enablers of antimicrobial stewardship. A summary of the major findings is provided (Table 8). The investigators also identified three major reasons for dispensing antimicrobials without a formal consultation in equine and cattle practice:[36]

1. Pressure to "keep clients happy" and fear that clients would be driven to competing practices if not provided with the desired antimicrobials.

2. Some clients felt they were able to diagnose common diseases and were unwilling to pay for a consultation for routine disease management.

3. Long work hours and lack of time made practitioners inclined to dispense the desired antimicrobial rather than spend time convincing clients of the importance of formal consultation or the lack of need for antimicrobials.

Table 8. Summary of major barriers and enablers for implementing antimicrobial stewardship programs in veterinary practices in Australia[36]

Major barriers	Major enablers
Client expectations and competition between practices	Concern for human health
Cost of microbiological testing	Pride in service provided
Lack of access to education and training	Low level of resistance encountered
Lack of antimicrobial stewardship governance structures	Preparedness to change prescribing practices
Lack of independent guidelines for antimicrobial use	Frequent use of low-cost diagnostic tests
Hierarchical structure of many practices	Low use of most critically important antimicrobial agents

The investigators noted that requests for antimicrobials without formal consultation was not reported by companion animal practitioners. However, it is the author's experience that exotic animal practitioners encounter these requests from time to time. Explanations for this behavior have included:

1. Ownership of a large collection physically limiting the owner's ability to present all affected patients

2. Recurrence of clinical signs for which the patient has been presented in the past

3. Development of clinical signs commonly encountered in the species and which frequently have an infectious component (e.g., dyspnea in rats, sneezing in rabbits, dermatitis in reptiles)

4. The perceived value of the pet is less than the cost of consultation

5. The veterinary practice caters almost exclusively towards individual pets, with no accommodations for herd medicine (e.g. no pricing for group/herd consultations, no equipment or predefined workflow for mobile consultations)

The author suggests that exotic animal practitioners accommodate herd medicine in their practices or develop a relationship with a mobile veterinary practice to which these clients may be referred.

Antibiograms—surveys of bacterial isolates and susceptibility patterns—are being more widely used in small animal medicine but are less commonly encountered in exotic pet practice. Published antibiograms vary from surveys of all cultures isolated from a class of patients (e.g., reptiles)[37,38] to summaries of results from specific lesions in one species[39,40]. Pathogen prevalence and susceptibility may vary widely not just by species and site of infection but also geographic area and client base. Ideally each practice would develop its own antibiogram in order to better understand the susceptibility patterns being encountered.[41] An antibiogram when combined with point-of-care cytology and Gram stain from a lesion can prove very useful in guiding drug selection pending culture and sensitivity results.

The field of exotic animal medicine currently suffers from a paucity of antimicrobial pharmacokinetic and pharmacodynamic data. Many drug dosing regimens employed in our field are based on anecdotal experience and perceived response. There is little doubt that this leads to inappropriate treatment and increased chances of resistance developing. The importance of continued determination of pharmacokinetic profiles in exotic species cannot be overemphasized.

Client Communication

Clients commonly present sick pets with the expectation of receiving antibiotics. This is especially true with exotic pets, which for years have been regarded by many veterinarians as having an inherent Baytril deficiency. It is essential to convey to pet owners the importance of proper antimicrobial use, as these drugs are not benign. The author frequently references the current worldwide public health crisis of antimicrobial resistance of the development of "superbugs," and often finds that clients have encountered this information before and understand the need for judicious use of these drugs.

The costs of diagnostics and therapy should be communicated upfront, even if the client says that the cost does not matter. It is especially important not to make assumptions about what the client will or will not pay based on their appearance or the species of pet (the author regularly encounters clients who are willing to fund thorough diagnostic investigations of their exotic pets). The costs should be placed in the context of the patient's health and prognosis, with the additional perspective of the potential impact on the client's health if antimicrobial resistance were to emerge in the patient. An estimate is only part of the discussion; a discussion with the owner must be had in order to put the recommendations in context, with a focus on outcome and well-being for the pet.[42]

Conclusion

Antimicrobial resistance is a towering problem, one that can only be solved through all medical professions collaborating to achieve a high level of expertise in the prescription of antimicrobials. As a clinician, the refusal to just send home two weeks of Baytril is empowering. It requires the practitioner to investigate, think, and develop a level of expertise that most exotic pet owners welcome. They want to know what is wrong with their pet and how to fix it; they don't want you to shrug your shoulders and dispense medication that **might** help. We can do better, and there is a way forward.

Endnotes

a. AVMA Veterinary Checklist for Antimicrobial Stewardship: https://www/avma.org/system/files?file=2020-03/Veterinary-Checklist-Antimicrobial-Stewardship.pdf.

References

1. Centers for Disease Control and Prevention. Antibiotic resistance threats in the United States, 2019. Atlanta: US Dept of Health and Human Services, CDC; 2019. Available at: https://www.cdc.gov/drugresistance/pdf/threats-report/2019-ar-threats-report-508.pdf. Accessed 22 Mar 2020. DOI: 10.15620/cdc:82532.

2. McEwen SA, Collignon PJ. Antimicrobial resistance: a one health perspective. *Microbiol Spectrum*. 2017;6(2). DOI: 10.1128/microbiolspec.ARBA-0009-2017.

3. Williams MR, Stedtfeld RD, Guo X, *et al.* Antimicrobial resistance in the environment. *Water Environ Res.* 2016;88(10):1951–1967. DOI: 10.2175/106143016X14696400495974.

4. Poirel L, Rodrigeuz-Martinez JM, Mammeri H, *et al.* Origin of plasmid-mediated quinolone resistance determinant QnrA. *Antimicrob Agents Chemother.* 2005;49(8):3523–3525. DOI: 10.1128%2FAAC.49.8.3523-3525.2005.

5. Poirel L, Kämpfer P, Nordmann P. Chromosome-encoded ambler class A beta-lactamase of *Kluyvera georgiana*, a probable progenitor of a subgroup of CTX-M extended-spectrum beta-lactamases. *Antimicrob Agents Chemother.* 2002;46(12):4038–4040. DOI: 10.1128/AAC.46.12.4038-4040.2002.

6. Patel R, Piper K, Cockerill FR, *et al.* The biopesticide *Paenibacillus popilliae* has a vancomycin resistance gene cluster homologous to the enterococcal VanA vancomycin resistance gene cluster. *Antimicrob Agents Chemother.* 2000;44(3):705–709. DOI: 10.1128%2Faac.44.3.705-709.2000.

7. Martinez JL. The role of natural environments in the evolution of resistance traits in pathogenic bacteria. *Proc R Soc B.* 2009;276:2521–2530. DOI: 10.1098/rspb.2009.0320.

8. Andersson DI, Hughes D. Evolution of antibiotic resistance at non-lethal drug concentrations. Drug resist update. 2012;15:162–172. DOI: 10.1016/j.drup.2012.03.005.

9. Davies J, Davies D. Origins and evolution of antibiotic resistance. *Microbiol Mol Biol Rev.* 2010;74(3):417–433. DOI: 10.1128/MMBR.00016-10.

10. Bueno I, Williams-Nguyen J, Hwang H, *et al.* Systematic review: impact of point sources on antibiotic-resistant bacteria in the natural environment. *Zoonoses Public Health.* 2017:1–23. DOI: 10.1111/zph.12426.

11. Weese JS, Giguère S, Guardabassi L, *et al.* ACVIM consensus statement on therapeutic antimicrobial use in animals and antimicrobial resistance. *J Vet Intern Med.* 2015;29:487–498. DOI: 10.1111/jvim.12562.

12. Dutil L, Irwin R, Ng LK, *et al.* Ceftiofur resistance in *Salmonella enterica* serovar Heidelberg from chicken meat and humans, Canada. *Emerg Infect Dis.* 2010;16(1):48–54. DOI: 10.3201/eid1601.090729.

13. Bager F, Madsen M, Christiansen J, *et al.* Avoparcin used as a growth promoter is associated with the occurrence of vancomycin-resistant *Enterococcus faecium* on Danish poultry and pig farms. *Prev Vet Med.* 1997;31:95–112. DOI: 10.1016/s0167-5877(96)01119-1.

14. Wellehan JFX, Divers SJ. Bacteriology. In: Divers SJ, Stahl SJ, eds. *Mader's Reptile and Amphibian Medicine and Surgery*. 3rd ed. St. Louis, MO: Elsevier, 2019;235–246.

15. Zur G, Gurevich B, Elad D. Prior antimicrobial use as a risk factor for resistance in selected *Staphylococcus pseudintermedius* isolates from the skin and ears of dogs. *Vet Dermatol.* 2016;27:468–e125. DOI: 10.1111/vde.12382.

16. Boothe DM, Debavalya N. Impact of routine antimicrobial therapy on canine fecal *Escherichia coli* antimicrobial resistance: a pilot study. *Int J Appl Res Vet Med.* 2011;9(4):396–406.

17. Schmidt VM, Pinchbeck G, McIntyre KM, *et al.* Routine antibiotic therapy in dogs increases the detection of antimicrobial-resistant faecal *Escherichia coli*. *J Antimicrob Chemother.* 2018;73:3305–3316. DOI: 10.1093/jac/dky352.

18. Boothe DM. Principles of antimicrobial therapy. In: Boothe DM, ed. *Small Animal Clinical Pharmacology and Therapeutics.* 2nd ed. St. Louis, MO: Saunders; 2012:248–352.

19. Bemis DA, Johnson BH, Bryant MJ, *et al.* Isolation and identification of *Caviibacter abscessus* from cervical abscesses in a series of pet guinea pigs (*Cavia porcellus*). *J Vet Diagn Invest.* 2016;28(6)763–769. DOI: 10.1177/1040638716665660.

20. Perry SM, Mitchell MA. Antibiotic therapy. In: Divers SJ, Stahl SJ, eds. *Mader's Reptile and Amphibian Medicine and Surgery*. 3rd ed. St. Louis, MO: Elsevier, 2019;1139–1154.

21. Broens EM, van Geijlswijk IM. Prudent use of antimicrobials in exotic animal medicine. *Vet Clin North Am Exot Anim Pract.* 2018;21:341–353. DOI: 10.1016/j.cvex.2018.01.014.

22. Weese JS, Blondeau JM, Boothe DM, *et al.* Antimicrobial use guidelines for treatment of urinary tract disease in dogs and cats: antimicrobial guidelines working group of the international society for companion animal infectious diseases. *Vet Med Int.* 2011:263768. DOI: 10.4061/2011/263768.

23. Hillier A, Lloyd DH, Weese JS, *et al.* Guidelines for the diagnosis and antimicrobial therapy of canine superficial bacterial folliculitis (antimicrobial guidelines working group of the international society for companion animal infectious diseases). *Vet Dermatol.* 2014;25:163–e43. DOI: 10.1111/vde.12118.

24. Lappin MR, Blondeau J, Boothe D, *et al.* Antimicrobial use guidelines for treatment of respiratory tract disease in dogs and cats: antimicrobial guidelines working group of the international society for companion animal infectious diseases. *J Vet Intern Med.* 2017;31:279–294. DOI: 10.1111/jvim.14627.

25. Divers SJ, Sladakovic I, Mäyer J, *et al.* Development of an antibiotic policy in a zoological medicine service and approach to antibiotic dosing using minimum inhibitory concentration data. In: Proceedings from the 2017 AEMV & ARAV; 2017:36–44.

26. World Health Organization. Critically Important Antimicrobials for Human Medicine. 5th revision. 2017. Available at: https://www.who.int/foodsafety/publications/antimicrobials-fifth/en/. Accessed 11 Apr 2020.

27. World Organization for Animal Health. OIE List of Antimicrobials of Veterinary Importance. 2019. Available at: https://www.oie.int/fileadmin/Home/eng/Our_scientific_expertise/docs/pdf/AMR/A_OIE_List_antimicrobials_July2019.pdf. Accessed 11 Apr 2020.

28. CLSI. Overview of factors affecting antimicrobial agent selection in animals. In: Understanding Susceptibility Test Data as a Component of Antimicrobial Stewardship in Veterinary

Settings. 1st ed. CLSI report VET09. Wayne: Clinical & Laboratory Standards Institute. 2019;10–28.

29. Petritz OA, Chen S. Therapeutic contraindications in exotic pets. *Vet Clin North Am Exot Anim Pract*. 2018;21:327–340.

30. Perry SM, Mitchell MA. Routes of administration. In: Divers SJ, Stahl SJ, eds. *Mader's Reptile and Amphibian Medicine and Surgery*. 3rd ed. St. Louis, MO: Elsevier; 2019;1130–1138.

31. Coutant T, Vergnau-Grosset C, Langlois I. Overview of drug delivery methods in exotics, including their anatomic and physiologic considerations. *Vet Clin North Am Exot Anim Pract*. 2018;21:215–259.

32. Prescott JF. Veterinary antimicrobial stewardship in North America. *Aust Vet J*. 2019;97(7):243–248.

33. American Veterinary Medical Association. Antimicrobial stewardship definition and core principles. Available at: https://www.avma.org/policies/antimicrobial-stewardship-definition-and-core-principles. Accessed 1 May 2020.

34. American Veterinary Medical Association. Judicious use of therapeutic antimicrobials. Available at: https://www.avma.org/policies/judicious-therapeutic-use-antimicrobials. Accessed 04 May 2020.

35. Davey P, Marwick CA, Scott CL, *et al*. Interventions to improve antibiotic prescribing practices for hospital inpatients. Cochrane Database Syst Rev 2017;2; CD003543. DOI: 10.1002/14651858.CD003543.pub4.

36. Hardefeldt LY, Gilkerson JR, Billman-Jacobe H, *et al*. Barriers to and enablers of implementing antimicrobial stewardship programs in veterinary practices. *J Vet Intern Med*. 2018;32:1092–1099. DOI: 10.1111/jvim.15083.

37. Cushing A, Pinborough M, Stanford M. Review of bacterial and fungal culture and sensitivity results from reptilian samples submitted to a UK laboratory. *Vet Rec*. 2011;169:390. DOI: 10.1136/vr.d4636.

38. Tang PK, Divers SJ, Sanchez S. Antimicrobial susceptibility patterns for aerobic bacteria isolated from reptilian samples submitted to a veterinary diagnostic laboratory: 129 cases (2005–2016). *J Am Vet Med Assoc*. 2020;257(3):305–312. DOI: 10.2460/javma.257.3.305.

39. Miranikova A, Hauptman R, Knotek Z, *et al*. Microbial flora of odontogenic abscesses in pet guinea pigs. *Vet Rec*. 2016;179(13):331. DOI: 10.1136/vr.103551.

40. Gardhouse S, Guzman DSM, Paul-Murphy J, *et al*. Bacterial isolates and antimicrobial susceptibilities from odontogenic abscesses in rabbits: 48 cases. *Vet Rec*. 2017;181(20):538. DOI: 10.1136/vr.103996.

41. Fowler H, Davis MA, Perkins A, *et al*. Survey of veterinary antimicrobial prescribing practices, Washington State 2015. *Vet Rec*. 2016;179(25):651. DOI: 10.1136/vr.103916.

42. Coe JB, Adams CL, Bonnett BN. A focus group study of veterinarians' and pet owners' perceptions of the monetary aspects of veterinary care. *J Am Vet Med Assoc*. 2007;231(10):1510–1518. DOI: 10.2460/javma.231.10.1510.

SPEAKER INFORMATION
(click the speaker's name to view other papers and abstracts submitted by this speaker)

Andrew Bean, DVM, MPH, DABVP (Exotic Companion Mammal Practice)
Avian & Exotic Medicine Service
Animal Emergency & Referral Center of Minnesota
Oakdale, MN, USA

URL: https://www.vin.com/doc/?id=9768912

Practical Approach to Rabbit Anesthesia

ExoticsCon Virtual 2020 Proceedings

Nicola Di Girolamo[1], DMV, MSc (EBHC), GPCert (ExAP), PhD, DECZM (Herpetology); Stefania Scarabelli[2], DMV, DECVAA, DACVAA

[1]Zoological Medicine Service, Center for Veterinary Health Sciences, Oklahoma State University, Stillwater, OK, USA; [2]Clinica Veterinaria Modena Sud, Spilamberto, Italy, and Clinica Veterinaria Malpensa, Samarate, Italy

Abstract

Rabbit anesthesia is drastically different than anesthesia of domestic carnivores. Rabbits have a strikingly higher perianesthetic mortality as compared to dogs and cats. Based on a large study including 8209 rabbit anesthetic procedures, the risks of anesthetic and sedation-related death in rabbits was 1.39% (95% CI 1.14–1.64%) within 48 hours of the procedure, almost ten times greater than dogs. This indicated that approximately 1 rabbit out of 72 anesthetized or sedated died. The use of proper sedation protocols, multimodal analgesia including constant-rate infusions, induction without unnecessarily high anesthetic gas concentrations, and endoscopic-guided endotracheal intubation have drastically reduced the perianesthetic mortality in the authors' institutions. In the current presentation we are going to discuss available protocols and best practices to minimize the risk of anesthetic-related mortalities in rabbits.

Introduction

Rabbit anesthesia is drastically different than anesthesia of domestic carnivores. Rabbits have a strikingly higher perianesthetic mortality as compared to dogs and cats.[1,2] Based on a large study including 8209 rabbit anesthetic procedures, the risks of anesthetic and sedation-related death in rabbits was 1.39% (95% CI 1.14–1.64%) within 48 hours of the procedure, almost ten times greater than dogs.[1] This indicated that approximately 1 rabbit out of 72 anesthetized or sedated died. When stratifying rabbits by healthy/diseased health status, healthy rabbits had a 0.73% risk of mortality, and diseased rabbits had a 7.4% risk of mortality.[1] A more recent study found an even higher mortality rate within 72 hours of the procedure, equal to 4.8% (95% CI 0.025–0.086; 6 rabbits over 210 procedures).[2] Unfortunately, while these studies are useful in order to identify this risk and to describe the risk to owners, there is limited clinical information that can be extrapolated reliably from the published reports to guide and improve future anesthetic events.

A further difficulty for selecting an ideal rabbit anesthesia relies in the fact that most published research studies include populations limited in numbers and procedures, making difficult to generalize their results to clinical patients. In the current presentation we are going to discuss available protocols and best practices to minimize the risk of anesthetic-related mortalities in rabbits. The veterinary practitioner should keep in mind that multimodal analgesia is fundamental in rabbits as well as in any other living beings.

Preparation of the Patient

Intravenous Catheter Placement

Placement of an intravenous catheter should be performed in every anesthetic event in order to administer fluids, analgesic drugs and, above all, emergency drugs if required. The marginal ear vein is the preferred site, but the cephalic and saphenous veins can be used as well. It is authors' opinion that, in healthy animals, it is better to apply the intravenous catheter after premedication in order to avoid stress to the animal. If the patient is more critical and IM premedication is not appropriate, a racemic mixture of lidocaine and prilocaine (EMLA cream) can be applied 30 minutes before venipuncture in order to reduce pain and discomfort. If a catheter is already in place, it is important to check patency before anesthesia. It is key to perform intubation only once the rabbit has been properly premedicated and does not show any chewing or tongue movements. Preanesthetic medications are discussed later on in these proceedings.

Intubation

Fasting is avoided in rabbits, but it is fundamental to clean the rabbit's mouth before intubation in order to remove food residues that could be otherwise pushed inside the trachea and inhaled. In order to achieve endotracheal intubation (ETI), a suitable depth of anesthesia should be induced by means of premedication (including combinations of benzodiazepine, dissociative agents, opioids and/or alpha-2 agonists) and, if needed, induction agents (e.g., propofol or alfaxalone). Lidocaine (spray or the injectable preparation instilled topically) can be used to desensitize the larynx.

Regardless of the method used for ETI, the rabbit should be placed in sternal recumbency with the neck and head extended dorsally and cranially but not completely vertical. A 2–4.4 mm ID uncuffed ETT should be available.

ETT in rabbits can be performed in different ways:

- **Endoscopically guided ETI**: A 2.7 mm endoscope at 30 or 0 degrees can be used. If feasible, it can be inserted into the endotracheal tube (ETT) and used as a guide to visualize the glottis and as a stylet to facilitate the ETT insertion. In ETT of smaller diameters a "side by side" technique is used, just to simply visualize the glottis. Endoscopic intubation is the only technique currently in use by the authors because (1) it allows visualization of the glottis and therefore inappropriate transport of food material into the trachea with the ETT; (2) once mastered, it is fast and allows adaptation to the different type of larynx of rabbits; and (3) it allows identification of alterations at the level of the glottis (e.g., laryngeal neoplasia).
- **Blind**: The ETT is inserted into the mouth and the breath sounds of the rabbit or the presence of condensation on the ETT itself are used to guide the insertion.
- **Otoscope**: For direct visualization of the larynx. The otoscope is inserted into the side of the patient's mouth, then moved centrally and used to elevate the soft palate and visualize the glottis. An ETT can be inserted through the cone of the otoscope or with a "side by side" technique.

In the authors' experience, endoscopically guided ETI allows insertion of a larger diameter ETT, thus decreasing airway resistance and work of breathing for the rabbits and allowing more accurate and reliable monitoring of end-tidal carbon dioxide ($EtCO_2$). Since a few years ago, a supraglottic airway device (V-GEL, Docsinnovent) has been produced and marketed for use in rabbits. The device is made of a soft gel-like material and should provide a pressure seal around the airway and the esophagus. The authors had instances in which V-GELs resulted in poor rabbit oxygenation (low SpO_2) and dislocated from the front of the glottis of the rabbit, resulting in ventilation in the gastrointestinal tract. Until a properly sized, clinical-based study compares the safety and effectiveness of supraglottic airway devices in rabbits with standard endotracheal intubation, the authors' advice is to maintain the use of endotracheal intubation.

Induction of General Anesthesia (Isoflurane, Sevoflurane, Propofol and Alfaxalone)

Historically, induction with anesthetic gases has been considered safer in exotic small mammals than induction with injectable agents (such as propofol, etomidate or alfaxalone). However, there are several shortcomings associated with induction of anesthesia with anesthetic gases that need to be considered.

Isoflurane in rabbits causes apnea during induction and a severe cardiovascular depression at high concentrations.[3,4,5] These are two main reasons why the use of anesthetic gases and especially the widely used isoflurane should be tailored in rabbits. Considering anesthetic mortality in healthy patients, it is possible that isoflurane accounts for some of the deaths observed. When isoflurane was administered in rabbits for anesthetic induction, long periods of apnea (>1 min) occurred together with bradycardia, hypercapnia and hypoxaemia.[5] In another study, most animals struggled violently during induction with both isoflurane and sevoflurane.[3] The authors of these studies warned against the use of isoflurane, sevoflurane, and desflurane for anesthetic induction of rabbits.[3,5]

During anesthesia, reaching high isoflurane concentrations in rabbits results in a dose-dependent cardiopulmonary depression attributable to vasodilation and negative inotropy.[4] At an isoflurane concentration of 4.15% with mechanical ventilation, cardiovascular depression is severe.[4] The authors mention that use of unnecessarily high isoflurane concentrations in this species should be avoided.[4]

The widespread use of endoscopic intubation has allowed a more consistent use of injectable agents for induction. The window of time required to intubate the rabbit may be shorter as compared to anesthetic gases (i.e., the rabbit may need to be intubated more quickly), but there are several cardiovascular benefits associated with the use of different injectable agents as induction.

Amongst injectable agents, propofol is commonly used, above all in rabbits, but recently alfaxalone has been licensed in this species as well. The main advantage of alfaxalone, compared with propofol, is that it can be used also IM (even if it is not licensed in rabbits for this route yet). When used IV, a dose-dependent respiratory depression and apnea are commonly reported.[6,7] The Summary of Product Characteristics (SPC) of Alfaxan suggests that rabbits are pre-oxygenated before induction of general anesthesia and that oxygen is provided throughout the entire anesthetic procedure. Dosage is reported as 4 mg/kg in premedicated rabbits and 5 mg/kg in non-premedicated animals even if lower doses (1–2 mg/kg) are generally efficacious. Despite the fact that propofol is still the authors' agent of choice for induction of anesthesia in rabbits, it must be remembered that apnea is common; therefore, the clinician should be confident with endotracheal intubation. Respiratory depression is dose related and can be decreased with slow administration of the drug. Nevertheless, we discourage inducing rabbits with propofol if endoscopic-guided intubation cannot be performed.

Constant Rate Infusions

Constant rate infusions (CRI) may be extremely advantageous in rabbits. A fentanyl CRI may reduce isoflurane MAC by approximately 60% in New Zealand white rabbits.[8] Furthermore, in rabbits anesthetized with isoflurane, mean arterial blood pressure, systemic vascular resistance and cardiac output were significantly higher when different fentanyl concentrations were obtained through CRIs.[9]

Lidocaine CRI at 50 mcg/kg/min resulted in the reduction of isoflurane MAC by 10.5%, while a 100 mcg/kg/min CRI resulted in the reduction of MAC by 21.7%.[10] Lidocaine also induced significant decreases in arterial blood pressure and heart rate, but the variables were still within the reference intervals for rabbits.[10] When compared to buprenorphine for management of postoperative pain, a CRI of lidocaine provided better postoperative outcomes with respect to fecal output, food intake, and glucose concentrations, with similar pain scores.[11] Thus, lidocaine appeared to be a suitable alternative to buprenorphine for alleviating postoperative pain with minimal risk of anorexia and gastrointestinal ileus.[11]

Vasopressors

Based on a recent study, phenylephrine might be more indicated than dopamine in order to increase MAP in anesthetized rabbits.[12] Dopamine increased stroke index at the highest infusion rate of 30 mcg/kg/min; however, cardiac output and mean arterial blood pressure remained unchanged from baseline values. Within the dose range of 5 to 30 mcg/kg/min, dopamine was not an effective treatment for isoflurane-induced hypotension in rabbits. Phenylephrine instead, at a rate of 2 mcg/kg/min, increased mean arterial blood pressure to 62 mm Hg from the baseline value of 45 mm Hg. This was a result of an increase in systemic vascular resistance with a concomitant decrease in heart rate and no change in cardiac output. Phenylephrine was not effective at the dose of 0.125, 0.25, 0.5, 1.0 mcg/kg/min and only minimally effective at a dose of 2 mcg/kg/min. Blood lactate concentration increased with time when rabbits received either treatment.[12]

Anesthetic Drugs Commonly Used in Rabbits

Several anesthetic protocols have been used and published in rabbits. Classes of drugs commonly employed include sedatives (midazolam, alpha-2 agonists, acepromazine), opioids (butorphanol, buprenorphine, fentanyl, methadone) and ketamine. These drugs should be used in different combinations and at different doses depending on the procedure performed, the required degree of sedation and analgesia and on the animal's temperament.

Alpha-2 agonists (medetomidine and dexmedetomidine) are commonly used in rabbits due to their sedative and analgesic properties and their anesthetic-sparing effect. However, important side effects must be remembered, above all at the cardiovascular level. Generally, there is an increase in systemic vascular resistance (SVR) and reflective reduction in heart rate followed by a centrally mediated bradycardia accompanied by normal or slightly increased/decreased SVR, depending on the species. In rabbits, SVR remains increased during dexmedetomidine infusions.[13] Regarding the respiratory system, dexmedetomidine has been reported to affect the respiratory response to carbon dioxide in rabbits.[14] The effects of alpha-2 agonists on gastrointestinal (GI) motility have been extensively studied both in human and veterinary medicine leading to controversial results. A recent study suggests a delayed transit motility in rabbits administered ketamine + medetomidine as compared to rabbits receiving ketamine + midazolam; however, the time to production of feces was identical in the two groups.[15] Other side effects of alpha-2 agonists include hyperglycaemia and enhanced diuresis. The authors generally use dexmedetomidine at doses of 25–35 mcg/kg IM + ketamine ± benzodiazepine ± opioids, either as a standalone anesthetic protocol for short and noninvasive procedures (e.g., dentals, castrations), or as a premedication before induction with propofol as needed, with intubation and maintenance on isoflurane or sevoflurane.

Benzodiazepines, particularly midazolam, can be used in rabbits. Midazolam can be given intramuscularly, and side effects are very limited. In the authors' experience, midazolam can be added to alpha-2 agonists, in very nervous healthy animals or can be given as the only sedative in critically ill patients. 0.5–1 mg/kg IM is the dose commonly used by the authors.

Ketamine can be used either for premedication or for induction of general anesthesia. Immobilization and analgesia are the main advantages offered by ketamine, whereas poor muscle relaxation is the main disadvantage; the latter can be overcome by adding benzodiazepines or alpha-2 agonists. Several studies evaluated the use of ketamine in association with other drugs in rabbits.[16] An effective premedication consists of ketamine (4–5 mg/kg up to 10 mg/kg) with alpha-2 agonists and benzodiazepine all administered intramuscularly.

Opioids are the mainstay for perioperative analgesia in humans, dogs and cats; their use in rabbits has always been controversial due to their effects on GI motility. Opioid receptors are highly expressed at the GI level and endogenous endorphins activation of mu-opioid receptors has been reported to inhibit GI motility.[17]

In rabbits, morphine has been reported to inhibit the small intestine's contractility *in vitro*,[18] whereas buprenorphine is the opioid more commonly evaluated in clinical studies. Several studies have demonstrated a negative effect of buprenorphine on GI motility.[11,19,20]

The authors' drugs of choice are methadone at doses of 0.2–0.3 mg/kg SC or IM or hydromorphone at doses of 0.3–0.5 mg/kg SC or IM. In a recent study on healthy rabbits[20] methadone administration was associated with no negative effects on appetite and GI motility, compared to buprenorphine and hydromorphone.[20] Butorphanol 0.3–0.4 mg/kg is used to enhance the activity of an alpha-2 agonist or benzodiazepine when sedation is required for nonpainful procedures, such as radiography or computed tomography. Nonsteroidal anti-inflammatory drugs and locoregional anesthetic techniques should always be considered in order to provide multimodal analgesia.

References

1. Brodbelt DC, Blissitt KJ, Hammond RA, Neath PJ, Young LE, Pfeiffer DU, Wood JL. The risk of death: the confidential enquiry into perioperative small animal fatalities. *Vet Anaesth Analg*. 2008;35(5):365–373.

2. Lee HW, Machin H, Adami C. Peri-anaesthetic mortality and nonfatal gastrointestinal complications in pet rabbits: a retrospective study on 210 cases. *Vet Anaesth Analg*. 2018;45(4):520–528.

3. Flecknell PA, Roughan JV, Hedenqvist P. Induction of anaesthesia with sevoflurane and isoflurane in the rabbit. *Lab Anim*. 1999;33(1):41–46.

4. Barter LS, Epstein SE. Cardiopulmonary effects of three concentrations of isoflurane with or without mechanical ventilation and supramaximal noxious stimulation in New Zealand white rabbits. *Am J Vet Res*. 2013;74(10):1274–1280.

5. Hedenqvist P, Roughan JV, Antunes L, Orr H, Flecknell PA. Induction of anaesthesia with desflurane and isoflurane in the rabbit. *Lab Anim*. 2001;35(2):172–179.

6. Gil AG, Silván G, Villa A, Illera JC. Heart and respiratory rates and adrenal response to propofol or alfaxalone in rabbits. *Vet Rec.* 2012;170(17):444. doi:10.1136/vr.100573.

7. Grint NJ, Murison PJ. A comparison of ketamine-midazolam and ketamine-medetomidine combinations for induction of anaesthesia in rabbits. *Vet Anaesth Analg.* 2008;35(2):113–121.

8. Barter LS, Hawkins MG, Pypendop BH. Effects of fentanyl on isoflurane minimum alveolar concentration in New Zealand White rabbits (*Oryctolagus cuniculus*). *Am J Vet Res.* 2015;76(2):111–115. doi:10.2460/ajvr.76.2.111.

9. Tearney CC, Barter LS, Pypendop BH. Cardiovascular effects of equipotent doses of isoflurane alone and isoflurane plus fentanyl in New Zealand White rabbits (*Oryctolagus cuniculus*). *Am J Vet Res.* 2015;76(7):591–598.

10. Schnellbacher RW, Carpenter JW, Mason DE, KuKanich B, Beaufrère H, Boysen C. Effects of lidocaine administration via continuous rate infusion on the minimum alveolar concentration of isoflurane in New Zealand White rabbits (*Oryctolagus cuniculus*). *Am J Vet Res.* 2013;74(11):1377–1384.

11. Schnellbacher RW, Divers SJ, Comolli JR, *et al.* Effects of intravenous administration of lidocaine and buprenorphine on gastrointestinal tract motility and signs of pain in New Zealand White rabbits after ovariohysterectomy. *Am J Vet Res.* 2017;78(12):1359–1371.

12. Gosliga JM, Barter LS. Cardiovascular effects of dopamine hydrochloride and phenylephrine hydrochloride in healthy isoflurane-anesthetized New Zealand White rabbits (*Oryctolagus cuniculus*). *Am J Vet Res.* 2015;76(2):116–121.

13. Sazuka S, Matsuura N, Ichinohe T. Dexmedetomidine dose dependently decreases oral tissue blood flow during sevoflurane and propofol anesthesia in rabbits. *J Oral Maxillofac Surg.* 2012;70(8):1808–1814.

14. Chang C, Uchiyama A, Ma L, Mashimo T, Fujino Y. A comparison of the effects on respiratory carbon dioxide response, arterial blood pressure, and heart rate of dexmedetomidine, propofol, and midazolam in sevoflurane-anesthetized rabbits. *Anesth Analg.* 2009;109(1):84–89.

15. Botman J, Hontoir F, Gustin P, *et al.* Postanaesthetic effects of ketamine-midazolam and ketamine-medetomidine on gastrointestinal transit time in rabbits anaesthetised with isoflurane. *Vet Rec.* 2020;186(8):249.

16. Grint NJ, Murison PJ. A comparison of ketamine-midazolam and ketamine-medetomidine combinations for induction of anaesthesia in rabbits. *Vet Anaesth Analg.* 2008;35(2):113–121.

17. Müller-Lissner S, Bassotti G, Coffin B, *et al.* Opioid-induced constipation and bowel dysfunction: a clinical guideline. *Pain Med.* 2017;18(10):1837–1863.

18. Feng ZY, Mao LG, Lu Y. [Effect of morphine chloride on contractility of small intestinal muscle *in vitro* or *in vivo* and its mechanism]. *Zhejiang Da Xue Xue Bao Yi Xue Ban.* 2008;37; 271–275.

19. Martin-Flores M, Singh B, Walsh CA, Brooks EP, Taylor L, Mitchell LM. Effects of buprenorphine, methylnaltrexone, and their combination on gastrointestinal transit in healthy New Zealand White rabbits. *J Am Assoc Lab Anim Sci.* 2017;56(2):155–159.

20. Pathak D, Di Girolamo N, Maranville R, Womble W, Sypniewski L, Hanzlicek A, Brandão J. Effects of injectable analgesics on selected gastrointestinal physiological parameters in rabbits. *Proceedings ExoticsCon.* 2020.

SPEAKER INFORMATION
(click the speaker's name to view other papers and abstracts submitted by this speaker)

Stefania Scarabelli, DMV, DECVAA, DACVAA
Clinica Veterinaria Modena Sud
Spilamberto, Italy

Nicola Di Girolamo, DMV, MSc(EBHC), GPCert(ExAP), PhD, DECZM (Herpetology)
Zoological Medicine Service, Center for Veterinary Health Sciences
Oklahoma State University
Stillwater, OK, USA

African Pygmy Hedgehogs: Husbandry and Clinical Management

ExoticsCon Virtual 2020 Proceedings

Grayson Doss, DVM, DACZM

Department of Surgical Sciences, University of Wisconsin-Madison, School of Veterinary Medicine, Madison, WI, USA

Abstract

The African pygmy hedgehog (*Atelerix albiventris*) is a nocturnal insectivore, grouped in the order Eulipotyphla along with shrews and voles. African pygmy hedgehogs should be housed in well-ventilated enclosures and provided exercise wheels. Supplemental heat may be necessary to prevent pet hedgehogs from entering a torporous state. Diet consists of a high-protein, moderate-fat kibble as the base, along with additional supplemented food items like vegetables and live insects. Physical examination and diagnostic sample collection are challenging in awake hedgehogs due to an inclination to roll into a protective ball of keratinaceous spines. Chemical immobilization is necessary for a complete physical examination in hedgehogs. Common disorders include hemorrhagic vulvar discharge, neoplasia, parasitic dermatologic disease, neurologic disease and oral cavity disease.

Introduction

African pygmy hedgehogs (*Atelerix albiventris*) are popular pets due to their inquisitive, often convivial nature; however, because of their unique anatomy and behavior, hedgehogs can be challenging patients. The lack of prospective research in this species forces veterinarians to rely on anecdotal information, which could lead to difficulties with diagnostics and ineffective or even dangerous treatment strategies.

Natural History and Anatomy

The African pygmy or central African hedgehog (*Atelerix albiventris*) is a nocturnal insectivore originating from central and eastern Africa. Hedgehogs are family Erinaceidae and were previously classified in the now obsolete order Insectivora. They are now grouped in the order Eulipotyphla that includes other small mammals like shrews and moles. The two most familiar hedgehog species are the central African hedgehog (*Atelerix albiventris*) and the European hedgehog (*Erinaceus europaeus*).

African pygmy hedgehogs are very active invertebrate predators and travel long distances at night to search for food. They are territorial, solitary animals and can enter a state of torpor in the wild during harsh environmental conditions.

The mantle is the formidable covering of the crown (head) and dorsum consisting of thousands of keratinaceous spines. There are four major muscles that function to roll the hedgehog into an impenetrable ball of spikes[1]: the *caudodorsalis* m. and the *frontodorsalis* m. start the action by rolling up the spines covering the rump and head, respectively. The *panniculus carnosis* m. then rolls up the dorsum of the hedgehog and the *orbicularis* m. cinches the mantle closed around the animal, like a purse string. The mantle should be moved dorsally when performing lateral radiographic views of the thorax so as to not interfere with the lung fields. The mantle spines are not barbed and have a basal bulb connecting the spine to the follicle that makes removal of healthy spines difficult. Teeth are brachydont, and the first maxillary incisor teeth have a considerable gap between them. The tail is short and spineless.

Sexing is easily accomplished by examining the anogenital distance. In females, the vulvar opening is located immediately cranial to the anus. In males, the preputial opening is located mid-abdomen, a great distance from the anus. Males have dramatic accessory sex glands (particularly the seminal vesicles) which often fill a large portion of the caudodorsal abdomen and can be mistaken for an abdominal mass when palpating dorsal to the urinary bladder. The testicles are mobile and sit in a para-anal recess rather than a scrotal sac. Females have a tightly curled, bicornuate uterus that has a "ram's horns" appearance. The mesometrium is short and the ovaries are small and often hidden within adipose tissue.

Husbandry

Housing should be well-ventilated and cleaned regularly, as hedgehogs produce copious amounts of soft feces. A large, high-walled, smooth-sided plastic storage container is easy to move and sanitize. Because of their active nature and propensity for obesity in captivity, a solid surface exercise wheel should be provided and cleaned daily if frequently used, due to rapid buildup of feces. Hedgehogs quickly learn to drink from either standing water or bottles. Bedding used for hedgehog enclosures varies from aspen shavings to artificial fleece material. Recycled paper bedding is easily cleaned, produces minimal dust, and a thick layer provides hedgehogs with opportunities to dig and burrow.

Insectivorous mammals are traditionally fed diets that are high in protein and have moderate fat content. Commercial hedgehog kibble or a high-quality weight management dry cat food is often used for the base diet. Vegetables may also be offered. Hedgehogs should be supplemented with live, low-fat content invertebrates for enrichment.

Physical Examination, Chemical Restraint and Analgesia

Performing a complete physical examination without chemical restraint is impossible, given hedgehog's shy nature and propensity to curl into a protected ball. Stress level of the individual animal and poor habituation to handling probably play the largest roles as to why hedgehogs remained curled up. Even hedgehogs that are used to handling and rarely ball up for their owners may become shy and non-approachable in a strange environment. While held suspended, holding the hedgehog with its ventrum facing the ground and then tilting the balled-up hedgehog head forward may coax tractable hedgehogs to unroll. For more stubborn individuals, placing them in a shallow (<2 cm depth) dish of warm water will often cause them to rapidly unroll in order to try and exit the water. During this attempted escape from the water, a cursory abdominal palpation may be performed. Scruffing can be performed but is likely stressful. Inhalant gas anesthesia with isoflurane or sevoflurane is the most common method of chemical restraint in African pygmy hedgehogs. Induction is via use of a small anesthetic chamber and then via facemask. Intubation requires small endotracheal tubes (1–2 mm I.D.), and it is difficult to visualize the glottis in this species. In the author's experience, a small supraglottic airway device designed for rabbits (v-gel R1, Docsinnovent Ltd, London, UK) provides improved airway support during general anesthesia when compared to a facemask and permits positive pressure ventilation.

There are few studies examining injectable anesthetic agents in this species, although use of various combinations of benzodiazepines, alpha$_2$-agonists, opioids, and ketamine have been anecdotally reported.[2] Administration of midazolam (0.5 mg/kg SC, IM; West-ward, Eatontown, NJ, USA) and butorphanol (0.3–0.5 mg/kg SC, IM; Torbugesic, Fort Dodge Animal Health, New York, NY, USA) may result in mild sedation which may permit scruffing. A combination of alfaxalone (3–5 mg/kg) and midazolam (1 mg/kg) administered deep SC into the mantle results in heavy sedation to light anesthesia useful for examination or basic, non-painful procedures like diagnostic imaging. It is partially reversed with flumazenil (0.05 mg/kg, deep SC).

While there is little information regarding analgesia in hedgehogs, a recent study[3,4] determined that a single, deep SC dose of buprenorphine hydrochloride administered into the mantle at 0.03 and 0.05 mg/kg provided analgesia for 36 and 48 hours, respectively, without apparent side effects.

Common Conditions and Presentations

Neoplasia

Neoplasia occurs commonly in African pygmy hedgehogs and is well reported in the scientific literature. Major organ systems affected include integumentary, hemolymphatic and alimentary.[5] Oral squamous cell carcinoma is one of the most commonly reported neoplasias in African pygmy hedgehogs.[5,6] Often, more than one type of malignant neoplasia has been reported in a single individual at the time of presentation. There are no studied oncologic treatments in pet hedgehogs.

Hemorrhagic Vulvar Discharge

At the author's institution, pet hedgehogs are commonly presented for hemorrhagic vulvar discharge. Urinalysis and abdominal ultrasound are useful tools for differentiating between two major differentials for hemorrhagic vulvar discharge: cystitis and uterine disease. Antibiotic treatment for cystitis should be based on urine bacterial culture and sensitivity. Uterine disease, often affecting the endometrium, is commonly reported in the literature,[7,8] with clinical signs including vaginal bleeding, hematuria, and weight loss. Recommended treatment for uterine disease is ovariohysterectomy.

Acariasis

Infestation with *Caparinia tripilis* is common in pet hedgehogs and can result in one or a combination of signs including spine or quill loss, seborrhea, pruritus, or skin erythema. Diagnosis is through skin scrapings of affected areas. Both mite sexes have three long setae on the third pair of legs and there are trilobite projections on both sides of the posterior end of male mites.[9] Selamectin (15–30 mg/kg topically q 14–30 d; Revolution, Zoetis Inc., Kalamazoo, MI, USA) or ivermectin (0.2–0.4 mg/kg SC or topically q 14 d; Ivomec, Boehringer Ingelheim, St. Joseph, MO, USA) may be used for treatment.

Oral Disease

A complete oral examination under anesthesia is very important for this species. Gingival masses, gingival hyperplasia and recession, periodontitis, gingivitis, calculus formation, dental fractures and abscesses may be noted during an oral exam. Antimicrobial therapy may be used for cases of gingivitis.[10] Dental prophylaxis may be performed as in other carnivores.

Obesity

Obesity is a common problem in captive hedgehogs. Healthy hedgehogs should be able to roll up completely, without any portion of the ventrum protruding. Obese hedgehogs may have large axillary or rump fat deposits or a decreased ability to roll up. Treatment includes eliminating high-fat foods, rationing the main diet, and increasing exercise. Hiding food in the substrate or disseminating it throughout the enclosure substrate can help promote weight loss through foraging and exercise. Weight reduction should be gradual to help prevent hepatic lipidosis, with owners monitoring their pet's weight with an accurate scale.

Neurologic Signs

In the author's experience, pet hedgehogs are often presented for ataxia. Wobbly hedgehog disease is a disorder of the white matter of the central nervous system leading to often progressive neurologic signs. The etiology of the disease or effective treatment options remain unknown. While "wobbly hedgehog syndrome" is an attractive differential, the author commonly encounters patients that have advanced systemic disease (e.g., severe anemia or metabolic derangement secondary to neoplastic disease) and are severely weak and stumble while ambulating, resulting in an abnormal gait that may be misinterpreted as true ataxia. Hedgehogs inadvertently allowed to enter a state of torpor may also present in a similar fashion, although the prognosis is much more favorable. Depending on the home environment, supplemental heating is often beneficial in preventing torpor by keeping temperatures in a preferred optimum range of 75 to 85°F (24 to 29°C). In southern African hedgehogs (*Atelerix frontalis*), torpor occurred in most animals at environmental temperatures ≤68°F (20°C).[11]

Normal neurologic examination findings have not been reported for African pygmy hedgehogs, making interpretation of findings and neurolocalization in animals presenting for suspected neurologic disease difficult. Additionally, considering the normally defensive behavior of many hedgehogs, performing several portions of the neurologic exam may be challenging to impossible.

References

1. Heatley JJ. Hedgehogs. In: Mitchell MA, Tully Jr TN, eds. *Manual of Exotic Pet Practice.* St. Louis, MO: Saunders Elsevier; 2009:433–455.

2. Helmer P, Carpenter J. Hedgehogs. In: Carpenter J, ed. *Exotic Animal Formulary.* 5th ed. St. Louis, MO: Saunders Elsevier; 2018:443–458.

3. Doss GA, Mans, C. Antinociceptive efficacy and safety of subcutaneous buprenorphine hydrochloride in African pygmy hedgehogs (*Atelerix albiventris*). *J Am Vet Med Assoc.* 2020. In press.

4. Doss GA, Mans C. Efficacy and safety of buprenorphine in African pygmy hedgehogs (*Atelerix albiventris*). *ExoticsCon.* 2019;8.

5. Raymond JT, Garner MM. Spontaneous tumours in captive African hedgehogs (*Atelerix albiventris*): A retrospective study. *J Comp Pathol.* 2001;124:128–133.

6. Gardhouse S, Eshar D. Retrospective study of disease occurrence in captive African pygmy hedgehogs (*Atelerix albiventris*). *Isr J Vet Med.* 2015;70:32–36.

7. Chambers JK, Shiga T, Takimoto H, *et al.* Proliferative lesions of the endometrium of 50 four-toed hedgehogs (*Atelerix albiventris*). *Vet Pathol.* 2018;55:562–571.

8. Mikaelian I, Reavill DR. Spontaneous proliferative lesions and tumors of the uterus of captive African hedgehogs (*Atelerix albiventris*). *J Zoo Wildl Med.* 2004;35:216–220.

9. Kim DH, Oh DS, Ahn KS, Shin SS. An outbreak of *Caparinia tripilis* in a colony of African pygmy hedgehogs (*Atelerix albiventris*) from Korea. *Korean J Parasitol.* 2012;50:151.

10. Lennox AM, Miwa Y. Anatomy and disorders of the oral cavity of miscellaneous exotic companion mammals. *Vet Clin North Am Exot Anim Pract.* 2016;19:929–945.

11. Hallam SL, Mzilikazi N. Heterothermy in the southern African hedgehog, *Atelerix frontalis*. *J Comp Physiol B.* 2011;181:437–445.

Speaker Information
(click the speaker's name to view other papers and abstracts submitted by this speaker)

Grayson A. Doss, DVM, DACZM
School of Veterinary Medicine
University of Wisconsin-Madison
Madison, WI, USA

URL: https://www.vin.com/doc/?id=9768911

Clinical Endocrinology of Small Mammals

ExoticsCon Virtual 2020 Proceedings

João Brandão, LMV, MS, DECZM (Avian)
Department of Veterinary Clinical Sciences, College of Veterinary Medicine, Oklahoma State University, Stillwater, OK, USA

Abstract

Endocrinology is the science that studies the influence of hormones on the body. The endocrine system has a major role on normal function of the body. Many diseases can affect the endocrine system, from neoplastic processes to abnormal response to physiological stimuli. Although endocrine disorders are well described in humans and small animals, there is limited information on exotic mammals. This master class will focus on disease processes affecting the endocrine glands and/or systems that are relevant in exotic mammals. During this master class, we will further review endocrine physiology, diagnostic tests for the assessment of the endocrine tissue, and treatment options for specific conditions.

Introduction

Endocrinology is the science that studies the influence of hormones on the body function.[1] Endocrine glands are specialized tissues/cells that synthesize, store, and release their hormonal secretions directly into blood.[2] The primary function of the endocrine system is to maintain normal metabolic function through internal and external environmental variation; a balance achieved through secretion of different hormones, some antagonistic, others synergistic.[3] Endocrine disorders appear to be involved in ~10 to 20% of canine and approximately 5% of feline disease processes.[4] In exotic animals there is relatively little information, but it is a growing field.

Diseases of the Pituitary Gland

Diabetes Insipidus (DI)

DI occurs due to failure to produce or secrete, or lack of renal effects of antidiuretic hormone. DI is further categorized as primary (neurogenic or central), nephrogenic, dipsogenic, or gestational.

Spontaneous DI in exotic small mammal species commonly kept as pets (rats and rabbits) can occur.[5-7] Diagnostic tests similar to those described for small animals are also adequate for exotic mammals. The use of the water deprivation test for diagnosing psychogenic polydipsia was reported in a laboratory New Zealand white rabbit (*Oryctolagus cuniculus*).[8] It is important to note that gerbils (*Meriones unguiculatus*) may become polydipsic (without DI) if food is restricted.[9]

Neoplasia

Pituitary neoplasms have been reported in multiple species of small exotic mammals (e.g., rabbits, ferrets), but the most prevalent species is rats.[10-12] Older female rats are most commonly diagnosed with pituitary neoplasms, and as a consequence, it has been suggested that estrogen may play a role in the disease.[13]

Reports of pituitary adenomas in pet rats and treatment of clinical cases are rare. A 2-year-old intact male albino pet rat was presented with a 3-week history of hypodipsia, suspected blindness, and behavioral changes.[14] The rat was subsequently treated with cabergoline (0.6 mg/kg PO q 72 h).[14] Clinical signs recurred at 8.5 months, and a pituitary adenoma was diagnosed on necropsy.[14]

Diseases of the Thyroid Gland

Clinical Assessment

The most common initial assessment of thyroid function is circulating serum thyroid hormones. Total thyroxine (TT_4) test has been shown to have poor sensitivity and specificity for thyroid disease.[15] In human medicine, TT_4 is not commonly used because the measurement of TT_4 and total triiodothyronine (TT_3) is unreliable since other nonthyroidal diseases can influence these values. This condition is commonly called euthyroid sick syndrome or nonthyroidal illness syndrome.[16] Euthyroid sick syndrome refers to changes in serum TSH, serum thyroid hormones, and tissue thyroid hormone levels that occur in patients with various nonthyroidal conditions; however, it is not a primary thyroid disease.[17] In cases of mild illness, only TT_3 decreases; however, in severe cases, both TT_3 and TT_4 decrease.[18]

Because of the unreliability of the total thyroid hormone measurement, this method has been largely abandoned in human medicine. Measurement of the free hormones, when measured by reliable methods, is more adequate. Furthermore, in cases of euthyroid sick syndrome or nonthyroidal illness syndrome, decreases in free thyroxine (fT_4) and free triiodothyronine (fT_3) are usually modest when compared to decreases in TT_3 and TT_4.[19] Nonetheless, free thyroid values can also be influenced by other nonthyroidal factors.[20]

Serum TSH measurement in humans is the single most reliable test to diagnose all common forms of hypothyroidism and hyperthyroidism.[21] Unfortunately, this diagnostic test is not validated for non-traditional species.

Another thyroid function diagnostic test, mainly for hypothyroidism, is dynamic testing. Dynamic testing implies administering a certain hormone and determining the physiologic response. Among exotic species, limited studies assessing dynamic thyroidal testing have been published. TRH and TSH stimulation testing have been reported in multiple mammalian species (e.g., rats, mice, ferrets, guinea pigs).[22-24]

Diagnostic imaging (e.g., nuclear medicine) has long been used for the diagnosis of thyroid dysfunction, specifically cases of hyperthyroidism. Taking advantage of the physiologic iodine/iodide cycle, an isotope with high affinity to the thyroid gland is administered, and uptake by the thyroid gland is either quantified or compared with another organ. Normal ratios have been formulated for small animals and, when elevated, may be indicative of a hyperfunctioning thyroid gland, supporting a possible hyperthyroid diagnosis. Reference intervals of thyroid scintigraphy for exotic species have not been well established, although rabbits have been reported in a non-peer review format.[25-27] Other imaging modalities used for thyroid diagnostic testing (e.g., ultrasonography, CT, and MRI) are mainly used for the diagnosis of neoplasia. Furthermore, these tests generally do not provide information regarding functionality.

Thyroid Dysfunction

Guinea Pigs

Guinea pigs (*Cavia porcellus*) are commonly diagnosed with hyperthyroidism. In addition to clinical reports, thyroid neoplasias have been reported as some of the most common neoplasias (3.6%) detected in guinea pigs by one laboratory service.[28] Age and gender predisposition have not been determined, but it appears that there may be a female bias and a higher predisposition for animals older than 3 years.[29] Several publications have reported normal thyroid reference intervals for guinea pigs, with specific intervals per sex, age, and environment.[30-34] Conversely, hypothyroidism is rarely diagnosed in small mammal species. There is only one limited report of hypothyroidism in a guinea pig, and the case description is nonspecific.[29]

Rabbits

Although rabbits are one of the most commonly kept exotic small mammal pets and have long been used in medical research, naturally occurring and induced thyroid dysfunction are rarely reported.[35-38] Although rabbit clinical thyroid disorders have not been published, the author is aware of several highly suspected cases of thyroid dysfunction. Thyroid dysfunction (e.g., euthyroid sick syndrome or nonthyroidal illness) was induced in rabbits that suffered skin wounds through experimental thermal means and has been reported in clinical cases.[37,39-41]

Chinchillas

One case of concurrent diabetes mellitus and hyperthyroidism has been reported in a chinchilla (*Chinchilla lanigera*).[42] Although correlation between the two conditions is considered unlikely, hyperthyroidism in cats can impair glucose tolerance, potentially leading to peripheral insulin resistance.[42,43]

Ferrets

Ferret hypothyroidism has been diagnosed in seven cases based on low basal TT_4 values and a limited response after TSH stimulation test.[44] The use of recombinant hTSH for stimulation testing has been validated in the ferret. Hypothyroid ferrets had similar clinical signs to other endocrinopathies: obesity, lethargy, inactivity, and excessive sleeping behavior.[44] There are at least two reports of thyroid neoplasias identified in ferrets: a C-cell carcinoma (medullary thyroid carcinoma) associated with multiple endocrine neoplasms and a thyroid follicular adenocarcinoma.[45,46]

Rats

Thyroid neoplasia, euthyroid sick syndrome or nonthyroidal illness, hypothyroidism, and hyperthyroidism have been reported in rats.[47,48]

Diseases Affecting the Endocrine Pancreas

Neoplasia

Insulinoma is an endocrinologically active insulin-secreting tumor of the pancreatic β cells. Hyperinsulinemia causes hypoglycemia by increased uptake of glucose by muscle tissue, fat, and the liver (all insulin target cells) and suppression of hepatic glycogenolysis and gluconeogenesis.[49] Neoplastic hyperfunctioning islet cells continue to produce insulin due to lack of negative feedback.[49]

Ferrets

Insulinomas are considered the most common reported neoplasias of ferrets, with an incidence of ~25% of all neoplasms diagnosed.[50] Insulinomas in ferrets tend to be benign and nonmetastatic, although local metastases have been suggested.[50]

Rats

Both spontaneous and induced insulinomas have been reported in laboratory rats.[51,52] In late stages of the disease, rats exhibited clinical signs of hypoglycemia, tachypnea, and partial limb paralysis.[53] The prevalence of spontaneous islet cell adenoma in aged laboratory rats appears to be low, with pancreatic islet cell adenomas reported to represent 2 to 6% of all tumors.[54-56]

Guinea Pigs

There have been three case reports of insulinomas in guinea pigs.[57] Insulinomas do not appear to be a common clinical problem in guinea pigs, with only one case being diagnosed *in vivo*.

Rabbits

Insulinoma was diagnosed in one rabbit; a serendipitous diagnosis was obtained during a study assessing the usefulness of blood glucose measurement in pet rabbits.[58]

Diabetes Mellitus (DM)

Rodents

Some rodents, in particular members of suborder Hystricomorpha, have some unique insulin characteristics and therefore unique physiology associated with DM. Spontaneous DM in guinea pigs has been reported in a laboratory colony and in a pet guinea pig.[59,60] Degus may be one of the most common exotic small mammal species that develop spontaneous diabetes during obesity.[61] In a retrospective study of degu diseases, DM was reported in 12/300, with a majority of animals (10/12) being 2 years old or older.[62]

Rabbits

Rabbits are used for diabetes research, and cases of spontaneous DM in New Zealand white rabbits have been reported.[63] However, spontaneous DM appears to be rare in rabbits.

Ferrets

DM has been reported in ferrets, most commonly iatrogenic following surgical removal of an insulinoma.[64] This condition tends to be transient, with insulin and glucose levels normalizing over a period of time. However, normalization of insulin levels may be due to recurrence of the insulinoma. Ferrets with DM may nonetheless require insulin therapy depending on severity of illness and the length of time necessary for resolution. Spontaneous DM unrelated to pancreatectomy and/or spontaneous pancreatitis has been reported.[65]

Diseases of the Adrenal Gland

Hyperadrenocorticism

Ferrets

Ferret hyperadrenocorticism, commonly called adrenal gland disease, is arguably the most well-studied endocrinopathy of exotic small mammals. Ferret adrenal gland disease is a form of hyperadrenocorticism caused by adrenal tissue hypertrophy and/or neoplasm within the adrenal cortex that overproduces one or more steroid hormones (e.g., glucocorticoids, mineralocorticoids, androgens).[50] It is important to mention that this condition is not the same as Cushing's syndrome (or disease), which is characterized by elevated cortisol levels that result from an ACTH-secreting pituitary tumor or a cortisol-secreting adrenal tumor.[50] It has been shown that neutering is the predisposing factor related to the development of hyperadrenocorticism in ferrets, independent of the age at which it is performed.[66] Furthermore, pituitary gonadotropic hormones also appear to have an influence on the pathogenesis of hyperadrenocorticism in ferrets.[67]

Hyperestrogenism

Rabbits

Hypertestosteronism associated with adrenal neoplasia and hyperplasia, leading to increased sexual and aggressive behavior, has been reported in older neutered rabbits.[68] Normal rabbit adrenal glands secrete appreciable amounts of sex steroids.[68]

References

1. Norris DO. *Vertebrate Endocrinology*. 4th ed. San Diego: Academic Press; 2007.

2. Capen CC. Tumors of the endocrine glands. In: *Tumors in Domestic Animals*. Ames, IA: Iowa State Press; 2008:607–696.

3. Marty MS, Carney EW, Rowlands JC. Endocrine disruption: historical perspectives and its impact on the future of toxicology testing. *Toxicol Sci*. 2011;120(suppl 1):S93–S108.

4. Engelking L, Rebar AH. Metabolic and Endocrine Physiology. Jackson, WY: CRC Press; 2012.

5. Cheng K, Friesen H, Martin J. Neurophysin in rats with hereditary hypothalamic diabetes insipidus (Brattleboro strain). *Endocrinology*. 1972;90:1055–1063.

6. Kalimo H, Rinne U. Ultrastructural studies on the hypothalamic neurosecretory neurons of the rat. II. The hypothalamoneurohypophysial system in rats with hereditary hypothalamic diabetes insipidus. *Z Zellforsch Mikrosk Anat*. 1972;134(2):205–225.

7. Boorman GA, Bree MM. Diabetes insipidus syndrome in a rabbit. *J Am Vet Med Assoc*. 1969;155:1218–1220.

8. Potter MP, Borkowski GL. Apparent psychogenic polydipsia and secondary polyuria in laboratory-housed New Zealand White rabbits. *J Am Assoc Lab Anim Sci*. 1998;37:87–89.

9. Kutscher CL, Stillman RD, Weiss IP. Food-deprivation polydipsia in gerbils (*Meriones unguiculatus*). *Physiol Behav*. 1968;3:667–671.

10. Lipman NS, Zhao ZB, Andrutis KA, *et al*. Prolactin-secreting pituitary adenomas with mammary dysplasia in New Zealand White rabbits. *Lab Anim Sci*. 1994;44:114–120.

11. Sikoski P, Trybus J, Cline JM, *et al*. Cystic mammary adenocarcinoma associated with a prolactin-secreting pituitary adenoma in a New Zealand White rabbit (*Oryctolagus cuniculus*). *Comp Med*. 2008;58:297–300.

12. Schoemaker NJ, van der Hage MH, Flik G, *et al*. Morphology of the pituitary gland in ferrets (*Mustela putorius furo*) with hyperadrenocorticism. *J Comp Pathol*. 2004;130:255–265.

13. Richardson VCG. *Diseases of Small Domestic Rodents*. 2nd ed. Oxford, UK; Malden, MA: Blackwell Publishing; 2003.

14. Mayer J, Sato A, Kiupel M, *et al*. Extralabel use of cabergoline in the treatment of a pituitary adenoma in a rat. *J Am Vet Med Assoc*. 2011;239:656–660.

15. Haarburger D. Thyroid disease: thyroid function tests and interpretation. *Contin Med Ed*. 2012;30(7):241–243.

16. Chopra IJ. Euthyroid sick syndrome: is it a misnomer? *J Clin Endocrinol Metab*. 1997;82:329–334.

17. McDermott MT. *Endocrine Secrets*. 6th ed. Philadelphia, PA: Elsevier Health Sciences; 2013.

18. De Groot LJ. Dangerous dogmas in medicine: the nonthyroidal illness syndrome. *J Clin Endocrinol Metab*. 1999;84:151–164.

19. Warner MH, Beckett GJ. Mechanisms behind the non-thyroidal illness syndrome: an update. *J Endocrinol*. 2010;205:1–13.

20. Midgley JE, Moon CR, Wilkins TA. Validity of analog free thyroxin immunoassays. Part II. *Clin Chem*. 1987;33:2145–2152.

21. Ladenson PW, Singer PA, Ain KB, *et al*. American Thyroid Association guidelines for detection of thyroid dysfunction. *Arch Intern Med*. 2000;160:1573–1575.

22. Heard DJ, Collins B, Chen DL, *et al*. Thyroid and adrenal function tests in adult male ferrets. *Am J Vet Res*. 1990;51:32–35.

23. Colzani RM, Alex S, Fang SL, *et al*. The effect of recombinant human thyrotropin (rhTSH) on thyroid function in mice and rats. *Thyroid*. 1998;8:797–801.

24. Mayer J, Wagner R, Mitchell MA, *et al*. Use of recombinant human thyroid-stimulating hormone for evaluation of thyroid function in guinea pigs (*Cavia porcellus*). *J Am Vet Med Assoc*. 2013;242: 346–349.

25. Harms CA, Hoskinson JJ, Bruyette DS, *et al*. Technetium-99m and iodine-131 thyroid scintigraphy in normal and radiothyroidectomized cockatiels (*Nymphicus hollandicus*). *Vet Radiol Ultrasound*. 1994;35:473–478.

26. Brandao J, Ellison M, Beaufrere H, *et al*. Thyroid scintigraphy determined by technetium-99m pertechnetate in euthyroid New Zealand White rabbits (*Oryctolagus cuniculus*). *Proc Int Conf Avian Herpetol Exot Mammal Med*. 2015:313.

27. Brandao J, Ellison M, Beaufrere H, *et al*. Quantitative 99m-technetium pertechnetate thyroid scintigraphy in euthyroid New Zealand White rabbits (*Oryctolagus cuniculus*). *Proceedings of the American College of Veterinary Radiology Annual Conference*. St. Louis: American College of Veterinary Radiology; 2014.

28. Gibbons PM, Garner MM, Kiupel M. Morphological and immunohistochemical characterization of spontaneous thyroid gland neoplasms in guinea pigs (*Cavia porcellus*). *Vet Pathol*. 2012;50:334–342.

29. Mayer J, Wagner R, Taeymans O. Advanced diagnostic approaches and current management of thyroid pathologies in guinea pigs. *Vet Clin North Am Exot Anim Pract*. 2010;13:509–523.

30. Fredholm DV, Cagle LA, Johnston MS. Evaluation of precision and establishment of reference ranges for plasma thyroxine using a point-of-care analyzer in healthy guinea pigs (*Cavia porcellus*). *J Exot Pet Med*. 2012;21:87–93.

31. Muller K, Muller E, Klein R, *et al*. Serum thyroxine concentrations in clinically healthy pet guinea pigs (*Cavia porcellus*). *Vet Clin Pathol*. 2009;38:507–510.

32. Castro MI, Alex S, Young RA, *et al*. Total and free serum thyroid hormone concentrations in fetal and adult pregnant and nonpregnant guinea pigs. *Endocrinology*. 1986;118:533–537.

33. Anderson RR, Nixon DA, Akasha MA. Total and free thyroxine and triiodothyronine in blood serum of mammals. *Comp Biochem Physiol A Comp Physiol*. 1988;89:401–404.

34. Quimby FW. *The Clinical Chemistry of Laboratory Animals*. 2nd ed. Philadelphia, PA: Taylor & Francis; 1999.

35. Kilicarslan H, Bagcivan I, Yildirim MK, *et al*. Effect of hypothyroidism on the NO/cGMP pathway of corpus cavernosum in rabbits. *J Sex Med*. 2006;3:830–837.

36. Kowalczyk E, Urbanowicz J, Kopff M, *et al*. Elements of oxidation/ reduction balance in experimental hypothyroidism. *Endokrynol Pol*. 2011;62:220–223.

37. Mebis L, Debaveye Y, Ellger B, *et al*. Changes in the central component of the hypothalamus-pituitary-thyroid axis in a rabbit model of prolonged critical illness. *Crit Care*. 2009;13(5):R147.

38. Mustafa S, Al-Bader MD, Elgazzar AH, *et al*. Effect of hyperthermia on the function of thyroid gland. *Eur J Appl Physiol*. 2008;103: 285–288.

39. Mebis L, Eerdekens A, Güiza F, *et al.* Contribution of nutritional deficit to the pathogenesis of the nonthyroidal illness syndrome in critical illness: a rabbit model study. *Endocrinology.* 2012;153:973–984.

40. Weekers F, Van Herck E, Coopmans W, *et al.* A novel *in vivo* rabbit model of hypercatabolic critical illness reveals a biphasic neuroendocrine stress response. *Endocrinology.* 2002;143:764–774.

41. Thöle M, Brezina T, Fehr M, *et al.* Presumptive nonthyroidal illness syndrome in pet rabbits (*Oryctolagus cuniculus*). *J Exot Pet Med.* 2019;31:100–103.

42. Fritsche R, Simova-Curd S, Clauss M, *et al.* Hyperthyroidism in connection with suspected diabetes mellitus in a chinchilla (*Chinchilla laniger*). *Vet Rec.* 2008;163:454–456.

43. Hoenig M, Ferguson D. Impairment of glucose tolerance in hyperthyroid cats. *J Endocrinol.* 1989;121:249–251.

44. Wagner R. Hypothyroidism in ferrets. *Proc Assoc Exotic Mammal Vet Conf.* 2012;2012:29–31.

45. Wills TB, Bohn AA, Finch NP, *et al.* Thyroid follicular adenocarcinoma in a ferret. *Vet Clin Pathol.* 2005;34:405–408.

46. Fox JG, Dangler CA, Snyder SB, *et al.* C-cell carcinoma (medullary thyroid carcinoma) associated with multiple endocrine neoplasms in a ferret (*Mustela putorius*). *Vet Pathol.* 2000;37:278–282.

47. Thorson L. Thyroid diseases in rodent species. *Vet Clin North Am Exot Anim Pract.* 2014;17:51–67.

48. Bray GA, York DA. Thyroid function of genetically obese rats. *Endocrinology.* 1971;88:1095–1099.

49. Lester NV, Newell SM, Hill RC, *et al.* Scintigraphic diagnosis of insulinoma in a dog. *Vet Radiol Ultrasound.* 1999;40:174–178.

50. Chen S. Advanced diagnostic approaches and current medical management of insulinomas and adrenocortical disease in ferrets (*Mustela putorius furo*). *Vet Clin North Am Exot Anim Pract.* 2010;13:439–452.

51. Inoue C, Shiga K, Takasawa S, *et al.* Evolutionary conservation of the insulinoma gene rig and its possible function. *Proc Natl Acad Sci.* 1987;84:6659–6662.

52. Flatt PR, Tan KS, Bailey CJ, *et al.* Effects of transplantation and resection of a radiation-induced rat insulinoma on glucose homeostasis and the endocrine pancreas. *Br J Cancer.* 1986;54:685–692.

53. Grant CS. Gastrointestinal endocrine tumours. Insulinoma. *Baillieres Clin Gastroenterol.* 1996;10:645–671.

54. Walsh K, Poteracki J. Spontaneous neoplasms in control Wistar rats. *Toxicol Sci.* 1994;22:65–72.

55. Bomhard E, Karbe E, Loeser E. Spontaneous tumors of 2000 Wistar TNO/W.70 rats in two-year carcinogenicity studies. *J Environ Pathol Toxicol Oncol.* 1986;7:35–52.

56. Bomhard E. Frequency of spontaneous tumors in Wistar rats in 30-months studies. *Exp Toxicol Pathol.* 1992;44:381–392.

57. Hess LR, Ravich ML, Reavill DR. Diagnosis and treatment of an insulinoma in a guinea pig (*Cavia porcellus*). *J Am Vet Med Assoc.* 2013;242:522–526.

58. Harcourt-Brown F, Harcourt-Brown S. Clinical value of blood glucose measurement in pet rabbits. *Vet Rec.* 2012;170:674.

59. Vannevel J. Diabetes mellitus in a 3-year-old, intact, female guinea pig. *Can Vet J.* 1998;39:503.

60. Belis JA, Curley RM, Lang CM. Bladder dysfunction in the spontaneously diabetic male Abyssinian-Hartley guinea pig. *Pharmacology.* 1996;53:66–70.

61. Spear G, Caple M, Sutherland L. The pancreas in the degu. *Exp Mol Pathol.* 1984;40:295–310.

62. Jekl V, Hauptman K, Knotek Z. Diseases in pet degus: a retrospective study in 300 animals. *J Small Anim Pract.* 2011;52:107–112.

63. Roth S, Conaway H. Animal model of human disease. Spontaneous diabetes mellitus in the New Zealand White rabbit. *Am J Pathol.* 1982;109:359.

64. Rosenthal KL, Wire NR. Endocrine diseases. In: Quesenberry K, Carpenter JW, eds. *Ferrets, Rabbits, and Rodents: Clinical Medicine and Surgery.* 3rd ed. St. Louis, MO: Elsevier/Saunders; 2012:86–102.

65. Benoit-Biancamano MO, Morin M, Langlois I. Histopathologic lesions of diabetes mellitus in a domestic ferret. *Can Vet J.* 2005;46:895–897.

66. Shoemaker NJ, Schuurmans M, Moorman H, *et al.* Correlation between age at neutering and age at onset of hyperadrenocorticism in ferrets. *J Am Vet Med Assoc.* 2000;216:195–197.

67. Schoemaker NJ, Teerds KJ, Mol JA, *et al.* The role of luteinizing hormone in the pathogenesis of hyperadrenocorticism in neutered ferrets. *Mol Cell Endocrinol.* 2002;197:117–125.

68. Lennox AM, Chitty J. Adrenal neoplasia and hyperplasia as a cause of hypertestosteronism in two rabbits. *J Exot Pet Med.* 2006;15:56–58.

Speaker Information
(click the speaker's name to view other papers and abstracts submitted by this speaker)

João Brandão, LMV, MS, DECZM (Avian)
Department of Veterinary Clinical Sciences
College of Veterinary Medicine
Oklahoma State University, Stillwater, OK, USA

URL: https://www.vin.com/doc/?id=9768914

Computed Tomography Scanning in Rabbits

ExoticsCon Virtual 2020 Proceedings

Jenna Richardson

Notes for this session are not yet available.

SPEAKER INFORMATION
(click the speaker's name to view other papers and abstracts submitted by this speaker)

Jenna Richardson

URL: https://www.vin.com/doc/?id=9768915

Using Comics to Teach Rabbit Hematology

ExoticsCon Virtual 2020 Proceedings

Joerg Mayer, DVM, MS, DABVP (Exotic Companion Mammal Practice), DECZM (Small Mammal), DACZM
Department of Small Animal Medicine and Surgery, College of Veterinary Medicine, University of Georgia, Athens, GA, USA

Abstract

We already know from anatomy and physiology that the rabbit is not like other small mammals in many different ways. This trend also applies to the clinical pathology of the rabbit. The lecture will focus on how to interpret the rabbit complete blood count using comics as a learning aid. The rabbit immune system reacts to challenges very differently than other mammals and having a good understanding of the rabbit CBC will be a significant aid in providing excellent clinical care for these beloved pets.

Introduction

The task of successful science education in the age of multimedia, when attention spans continually decrease, is very challenging. In the author's opinion, the importance of telling stories and providing visual aids to deliver content has unfortunately not reached the same popularity in the teacher's world when compared to the preferences of their younger audiences. The use of stories to teach science is not a novel concept, and science education has been a subject of research for centuries.[1] The author is using comics to illustrate and highlight specific important points when interpreting the hematology results of the rabbit.

A Type of Teaching Technique

It is essential to realize that a compelling story cannot be told with just one picture. Therefore, it is crucial to recognize that the bloodwork of your patient is only one part of the story. Do not make the mistake of overinterpreting the status of the animal just based on one data set. Consider the bigger picture, examine the animal, look at the history, and look at the present and post data sets. However, sometimes the clinician is forced to limit the data collection to a single diagnostic test or procedure. To offer the best service, always recommend and document that you provided a variety of diagnostic tests that can provide the most information to evaluate the patient. Also, educate the owner that you might need to run repeated samples, for example, to monitor the response to a specific treatment. You have to eventually connect the dots to draw a picture, and it gets easier the more dots you have to work with.

For the sake of the lecture, we will focus on blood work as the data to evaluate. One of the most critical aspects of looking at bloodwork (especially chemistries) is to validate the reference one uses. Do go through the extra effort to figure out what is the population that was used to establish the reference range you are using. There are a large variety of 'normal' values published, which may differ significantly. The reader is encouraged to compare and contrast the sources of the published values.

The total volume of blood available for collection depends on the health status of the rabbit. If the rabbit is healthy, about 1% of the circulating blood volume can be considered available for collection. If the animal is diseased or if repeated bleeding events are needed, 0.5% might be safer. The circulating blood volume of the rabbit is approximately 55–70 mL/kg of body weight.[2] Common phlebotomy sites for the rabbit include the cephalic vein, saphenous vein, and the lateral ear vein. Because rabbit blood tends to clot quickly and the cells can lyse readily, the blood should be immediately placed in an anticoagulant. Slides need also be made immediately to avoid post-sample artifacts. Anticoagulants commonly used (EDTA and heparin) work well for rabbit blood.

Erythrocytes have a shorter lifespan (average ~50–60 days) when compared to other domesticated mammals.[3] Hence polychromasia is typical when compared to other larger mammals. So, the image that comes to mind when comparing the erythrocyte character on the blood smear of a rabbit might be a shopping mall where the erythrocytes are the people in the mall. Everyone can picture this and realizes that there is a significant age variation in the population when compared to a workplace, like a cubicle world of a larger company. The cubicle world could represent the erythrocyte population in a dog (much less polychromasia) as there is significantly less variation in the age of the employees when compared to the shopping mall attendees.

The morphology of rabbit **lymphocytes** is similar to that of other species. Lymphocytes have a large nucleus, that may be slightly indented, and a small amount of light blue cytoplasm. Although small lymphocytes predominate, large lymphocytes may be present. Reactive lymphocytes (immunocytes) are usually antigenically stimulated lymphocytes that are larger cells with a more intensely blue cytoplasm. In the comic world analogy, the lymphocyte would be represented by a "mall cop" who controls the area and looks for any abnormal activities.

Heterophils have a light purple, lobulated nucleus surrounded by cytoplasm containing diffuse, variably sized reddish granules. They are often misidentified as eosinophils by the untrained eye during the complete blood count. In the literature, they are sometimes referred to as "pseudo-eosinophils." Heterophilic granules are generally smaller than those of eosinophils and may not occupy all of the cytoplasm. Heterophils range from 10 to 15 µm in diameter. Minor heterophil degranulation may accompany the use of rapid Romanowsky-type stains such as Diff-Quick. Stain-induced degranulation is presumed to be a sequel of short fixation time (5 to 10 seconds) during staining. Degranulated cells will still resemble heterophils, but vacuoles will be present where the granules previously were located. Stain-induced degranulation should not be confused with toxic change in which the cytoplasm has a blue cast. In the comic world analogy, the heterophil would be represented by a "special police unit" who would be called to the area, if any abnormal or significant violent activity would be reported by the "mall cop."

In the **monocyte**, a few cytoplasmic vacuoles may be observed. Large, dark red granules have been described in the cytoplasm of some monocytes in association with nonspecific toxic changes of leucocytes. The monocytes usually appear in higher numbers if the initial immune response is insufficient to clear the problem. In the comic analogy, this would be illustrated by the "hostage negotiator" trying to deescalate the situation.

Eosinophils are slightly larger than heterophils and are 12 to 16 µm in diameter. The nucleus stains purple and often appears bilobed. Intensely acidophilic, round, cytoplasmic granules are present that are larger and more numerous than the granules in heterophils.

Basophils have a light purple, lobulated nucleus and dark purple to purple-black cytoplasmic granules. They are approximately the same size as heterophils. Rabbits usually have more basophils than other mammals. Up to 30% of leukocytes in healthy rabbits has been seen. However, the significance of this observation is unclear.[4]

Because these last two cell lines play a minor role in the evaluation of the rabbit CBC, there is no significant role assigned to any of these in the comic strip.

Additional Evaluation

One of the main observations in the complete blood count (CBC) of a sick rabbit is a decrease in the hematocrit (HCT).[4] When this observation is translated to our comic world, it would be depicted as a mall where there is a fire or something seriously going on, and most of the attendees (RBC) are leaving the area while some of the mall cops (lymphocytes) are coming to investigate and trying to clear the problem.

The lecture will use a few clinical cases and real data to illustrate the analogy with the comic strip "shopping mall" image to highlight the most important aspects of interpreting the changes in the rabbit CBC.

The "Cliff Notes" in the appendix at the end of this document are designed to be printed out and laminated to have as a reference when reviewing a blood smear in your clinic.

Example Cases Discussed in Lecture

Case 1. Normal rabbit

Doctor:	HERSEY	Species:	LAGOMORPH
Owner:	GRONDIN, KERRY	Breed:	RABBIT
Patient:	5170140	Sex:	MC
Patient I.D.:	OLLIE	Age:	1 yrs. / 2 mos.

General Hematology

	Results	Normal Ranges	
WBC:	11.900 x10û3/ul	5.000 - 12.000	x10û3/ul
RBC:	6.84 x10û6/ul	4.0 - 8.4	x10û6/ul
Hemoglobin:	14.0 g/dl	8.80 - 17.50	g/dl
Hematocrit:	43 %	30 - 50	%
MCV:	60.3 fl	58.00 - 66.50	fl
MCH:	20.5 pg	17.50 - 23.50	pg
MCHC:	32.6 g/dl	29.00 - 37.00	g/dl
RDW:	16.6 %		

Differential

	Results Absolute	Results Relative	Normal Ranges Absolute	Normal Ranges Relative
Seg N:	4.998 x10û3/ul	42 %	0.136 - 8.100 x10û3/ul	35 - 55 %
Lymphocytes:	6.545 x10û3/ul	55 % <H>	0.323 - 7.770 x10û3/ul	25 - 50 %
Monocytes:	0.357 x10û3/ul	3 %	0.019 - 2.060 x10û3/ul	2 - 10 %

Comments

Plasma appearance: Normal
RBC Morphology: No significant abnormalities
Platelets: Platelet clumps seen on smear. Actual platelet count and/or estimate may be affected.
From 10-25 platelets were seen per 100x field. This corresponds to an estimated platelet count of 200,000-500,000/ul.
WBC Exam: No significant abnormalities

Case 2. Stress leukogram

Doctor:	MAYER, JOERG	Species:	LAGOMORPH
Owner:	TONELLI	Breed:	RABBIT
Patient:	S130904	Sex:	MC
Patient I.D.:	SHELDON	Age:	3 yrs. / 7 mos.

General Hematology

	Results	Normal Ranges	
WBC:	5.300 x10û3/ul	5.000 - 12.000	x10û3/ul
RBC:	6.18 x10û6/ul	4.0 - 8.4	x10û6/ul
Hemoglobin:	13.7 g/dl	8.80 - 17.50	g/dl
Hematocrit:	40 %	30 - 50	%
MCV:	63.8 fl	58.00 - 66.50	fl
MCH:	22.2 pg	17.50 - 23.50	pg
MCHC:	34.3 g/dl	29.00 - 37.00	g/dl
RDW:	15.4 %		

Differential

	Results Absolute	Results Relative	Normal Ranges Absolute	Normal Ranges Relative
Seg N:	2.703 x10û3/ul	51 %	0.135 - 8.100 x10û3/ul	35 - 55 %
Lymphocytes:	2.014 x10û3/ul	38 %	0.323 - 7.770 x10û3/ul	25 - 50 %
Monocytes:	0.424 x10û3/ul	8 %	0.019 - 2.060 x10û3/ul	2 - 10 %
Eosinophils:	0.159 x10û3/ul	3 %	0.017 - 2.691 x10û3/ul	0 - 5 %

Comments

Plasma appearance: Normal
RBC Morphology: No significant abnormalities
Platelets: >25 platelets were seen per 100x field. This corresponds to an estimated platelet count >500,000/ul.
WBC Exam: No significant abnormalities

Case 3. Diseased rabbit

General Hematology

	Results			Normal Ranges	
WBC:	16.300	x10û3/ul	<H>	5.000 - 12.000	x10û3/ul
RBC:	4.50	x10û6/ul		4.0 - 8.4	x10û6/ul
Hemoglobin:	10.0	g/dl		8.80 - 17.50	g/dl
Hematocrit:	30	%		30 - 50	%
MCV:	65.9	fl		58.00 - 66.50	fl
MCH:	22.2	pg		17.50 - 23.50	pg
MCHC:	33.3	g/dl		29.00 - 37.00	g/dl
RDW:	15.3	%			

Differential

	Results Absolute	Results Relative	Normal Ranges Absolute	Normal Ranges Relative
Seg N:	13.366 x10û3/ul <H>	82 % <H>	0.136 - 8.100 x10û3/ul	35 - 55 %
Lymphocytes:	1.793 x10û3/ul	11 % <L>	0.323 - 7.770 x10û3/ul	25 - 50 %
Monocytes:	1.141 x10û3/ul	7 %	0.019 - 2.060 x10û3/ul	2 - 10 %

Comments

Plasma appearance: Normal
RBC Morphology: Slight poikilocytosis
Platelets: Platelet clumps seen on smear. Actual platelet count and/or estimate may be affected.
From 10-25 platelets were seen per 100x field. This corresponds to an estimated platelet count of 200,000-500,000/ul.
WBC Exam: Occasional reactive lymphocyte present

Case 4. Septic rabbit

General Hematology

	Results			Normal Ranges	
WBC:	4.240	x10û3/ul	<L>	5.000 - 12.000	x10û3/ul
RBC:	3.74	x10û6/ul	<L>	4.0 - 8.4	x10û6/ul
Hemoglobin:	7.6	g/dl	<L>	8.80 - 17.50	g/dl
Hematocrit:	24	%	<L>	30 - 50	%
MCV:	62.2	fl		58.00 - 66.50	fl
MCH:	20.3	pg		17.50 - 23.50	pg
MCHC:	31.7	g/dl		29.00 - 37.00	g/dl
RDW:	14.1	%			

Differential

	Results Absolute	Results Relative	Normal Ranges Absolute	Normal Ranges Relative
Seg N:	3.731 x10û3/ul	88 % <H>	0.136 - 8.100 x10û3/ul	35 - 55 %
Lymphocytes:	0.466 x10û3/ul	11 % <L>	0.323 - 7.770 x10û3/ul	25 - 50 %
Monocytes:	0.042 x10û3/ul	1 % <L>	0.019 - 2.060 x10û3/ul	2 - 10 %

Comments

Plasma appearance: Normal
RBC Morphology: No significant abnormalities
Platelets: Platelet clumps seen on smear. Actual platelet count and/or estimate may be affected.
From 10-25 platelets were seen per 100x field. This corresponds to an estimated platelet count of 200,000-500,000/ul.
WBC Exam: few reactive lymphocytes seen

Case 5. Break-down phase of the immune system

Doctor:	MAYER, JOERG	Species:	LAGOMORPH
Owner:	NI CHONAILL	Breed:	RABBIT
Patient:	5175013	Sex:	FS
Patient I.D.:	BINKEY	Age:	2 yrs. / 12 mos.

General Hematology

	Results			Normal Ranges	
WBC:	1.670	x10û3/ul	<L>	5.000 - 12.000	x10û3/ul
Corrected WBC:	1.184	x10û3/ul	<L>	5.000 - 12.000	x10û3/ul
RBC:	3.28	x10û6/ul	<L>	4.0 - 8.4	x10û6/ul
Hemoglobin:	8.2	g/dl	<L>	8.80 - 17.50	g/dl
Hematocrit:	23	%	<L>	30 - 50	%
MCV:	68.6	fl	<H>	58.00 - 66.50	fl
MCH:	25.0	pg	<H>	17.50 - 23.50	pg
MCHC:	35.7	g/dl		29.00 - 37.00	g/dl
RDW:	13.3	%			

Differential

	Results Absolute	Results Relative	Normal Ranges Absolute	Normal Ranges Relative
Seg N:	0.284 x10û3/ul	24 % <L>	0.136 - 8.100 x10û3/ul	35 - 55 %
Lymphocytes:	0.900 x10û3/ul	76 % <H>	0.323 - 7.770 x10û3/ul	25 - 50 %
NRBC/100 WBC:	41			

Comments

Plasma appearance: Normal
RBC Morphology: No significant abnormalities
Platelets: From 10-25 platelets were seen per 100x field. This corresponds to an estimated platelet count of 200,000-500,000/ul.
WBC Exam: Neutrophils/heterophils appear moderately to markedly toxic (hypogranular, some with toxic granulation).

Case 6. Thymic lymphoma

General Hematology

	Results		Normal Ranges	
WBC:	4.800	x10ù3/ul <L>	5.000 - 12.000	x10ù3/ul
RBC:	6.34	x10ù6/ul	4.0 - 8.4	x10ù6/ul
Hemoglobin:	12.6	g/dl	8.80 - 17.50	g/dl
Hematocrit:	38	%	30 - 50	%
MCV:	59.4	fl	58.00 - 66.50	fl
MCH:	19.9	pg	17.50 - 23.50	pg
MCHC:	33.2	g/dl	29.00 - 37.00	g/dl
RDW:	24.0	%		

Differential

	Results Absolute	Results Relative	Normal Ranges Absolute	Normal Ranges Relative
Seg N:	1.248 x10ù3/ul	26 % <L>	0.135 - 8.100 x10ù3/ul	35 - 55 %
Lymphocytes:	3.456 x10ù3/ul	72 % <H>	0.323 - 7.770 x10ù3/ul	25 - 50 %
Monocytes:	0.096 x10ù3/ul	2 %	0.019 - 2.060 x10ù3/ul	2 - 10 %

Comments

Plasma appearance:	Normal
RBC Morphology:	Slight to moderate poikilocytosis Small to moderate numbers of acanthocytes and schistocytes present
Platelets:	From 10-25 platelets were seen per 100x field. This corresponds to an estimated platelet count of 200,000-500,000/ul.
WBC Exam:	Many lymphocytes appear reactive. A few atypical lymphocytes are seen. These are medium-sized cells with dispersed chromatin and small amounts of pale blue cytoplasm.
General comments:	The atypical lymphocytes seen any be reactive cells, but a neoplastic origin cannot be ruled out considering the diagnosis of lymphoma in this patient.

Case 7. Leukemia

Patient I.D.: GUIN PIG Age: 5 yrs. / 4 mos.

General Hematology

	Results		Normal Ranges
WBC:	61.500	x10ù3/ul	
RBC:	4.95	x10ù6/ul	
Hemoglobin:	14.1	g/dl	
Hematocrit:	42	%	
MCV:	81.2	fl	
MCH:	28.5	pg	
MCHC:	33.6	g/dl	
RDW:	17.3	%	

Differential

	Results Absolute	Results Relative	Normal Ranges Absolute	Normal Ranges Relative
Seg N:	14.145 x10ù3/ul	23 %		
Lymphocytes:	43.665 x10ù3/ul	71 %		
Monocytes:	3.075 x10ù3/ul	5 %		
Basophils:	0.615 x10ù3/ul	1 %		

Comments

Plasma appearance:	Slight hemolysis
RBC Morphology:	No significant abnormalities
Platelets:	Platelet clumps seen on smear. Actual platelet count and/or estimate may be affected. From 10-25 platelets were seen per 100x field. This corresponds to an estimated platelet count of 200,000-500,000/ul.
WBC Exam:	The lymphocytes consist of mixture of small, medium and large lymphocytes, the latter with diameters of 12-12 microns. Medium to large lymphocytes account for approximately 35% of the leukocytes. The medium to large lymphocytes have nuclei that are round, indented or have slightly cerebriform nuclei lacking distinct nucleoli.

Hematocrit:	42	%
MCV:	81.2	fl
MCH:	28.5	pg
MCHC:	33.6	g/dl
RDW:	17.3	%

Differential

	Results Absolute	Results Relative	Normal Ranges Absolute	Normal Ranges Relative
Seg N:	14.145 x10û3/ul	23 %		
Lymphocytes:	43.665 x10û3/ul	71 %		
Monocytes:	3.075 x10û3/ul	5 %		
Basophils:	0.615 x10û3/ul	1 %		

Comments

Plasma appearance: Slight hemolysis

RBC Morphology: No significant abnormalities

Platelets: Platelet clumps seen on smear. Actual platelet count and/or estimate may be affected.
From 10-25 platelets were seen per 100x field. This corresponds to an estimated platelet count of 200,000-500,000/ul.

WBC Exam: The lymphocytes consist of mixture of small, medium and large lymphocytes, the latter with diameters of 12-12 microns. Medium to large lymphocytes account for approximately 35% of the leukocytes. The medium to large lymphocytes have nuclei that are round, indented or have slightly cerebriform nuclei lacking distinct nucleoli.

Foa-Kurloff bodies comprise from 3-6% of the leukocytes depending on the area of the smear examined. This may represent a slight increase in their numbers, although it has been reported that these cells can comprise as much as 3-4% of the differential count.

General comments: Heparinized sample received

Rabbit Hematology Cliff Notes

by Dr. Joerg Mayer

General points:

- See the bigger picture (always correlate findings to physical exam and other tests).
- Run repeat samples (connect the dots to see the trend).
- Always doubt (if the result does not make sense call the lab to verify).
- Pick reliable reference source and stick with that (reference values can vary significantly).
- For phlebotomy 1% of bwt can be safely taken (10 ml from a 1 kg animal).

Rabbit blood	Reference values
Clots very fast, fill EDTA first in case you fill more. If serum is needed, clot will be mature in about 30 minutes. The lymphocyte is the most common leukocyte in the blood of young animals.	WBC=6,300–10,060 cells/µl Heterophils=1,490–3,210 cells/µl Lymphocytes=3,360–7,000 cells/µl Monocytes=50–450 cells/µl Eosinophils=100–150 cells/µl Basophils=60–360 cells/µl

Changes in the sick rabbit	CBC in the septic rabbit
Many sick rabbits have hemoglobin and PCV much lower than normal count. Rabbits with a bacterial infection will not have leukocytosis but have increased heterophil count (absolute) and decrease in lymphocyte count (relative lymphopenia).	General hematology

	Results
WBC	4.240 x 10^3/µl <L>
RBC	3.74 x 10^6/µl <L>
Hemoglobin	7.6 g/dl <L>
Hematocrit	24%
MCV	62.2 fl
MCH	20.3 pg
MCHC	31.7 g/dl

RDU	14.1%

Neutropenia

Occurs when consumption is higher than production.
Often in acute inflammation or with marked severity of disease.

Differential

	Results	
	Absolute	Relative
Seg N	$3.731 \times 10^3/\mu l$	88%<L>
Lymphocytes	$0.466 \times 10^3/\mu l$	11% <L>
Monocytes	$0.042 \times 10^3/\mu l$	1% <L>

Summary

One of the main effects of disease is to reduce the total blood cellularity, particularly with respect to the red cell values but also to a lesser extent with the white cell counts.
The hematocrit might be one of the best indicators to differentiate between sick and healthy.
The least important variable was the basophil count, followed by the monocyte and eosinophil counts.

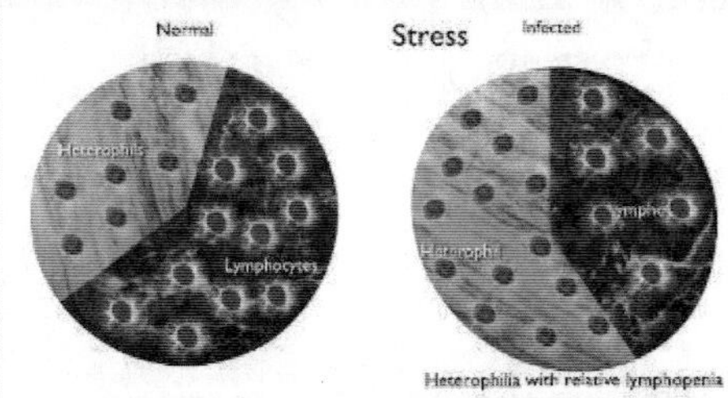

Lymphocyte

Reactive lymphocytes (immunocytes) are antigenically-stimulated lymphocytes that are larger cells with a more intensely blue cytoplasm.
The morphology of the rabbit lymphocytes is similar to that of other species.
Lymphocytes have a large nucleus, that may be slightly indented, and a small amount of light blue cytoplasm.
Although small lymphocytes predominate, large lymphocytes may be present.

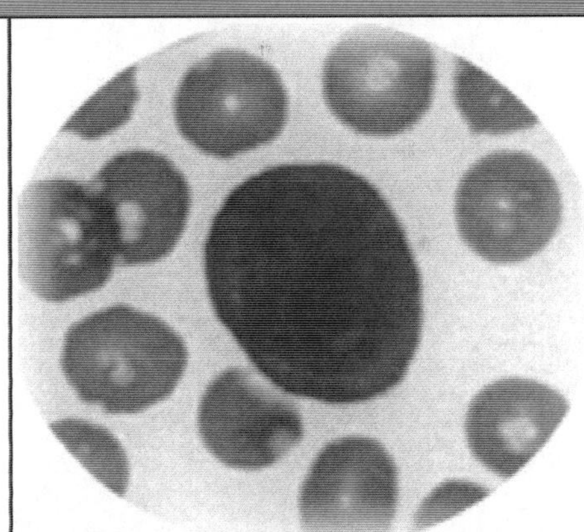

Heterophil

They have a light purple, lobulated nucleus surrounded by cytoplasm containing diffuse, variably sized reddish granules.
Often misidentified as eosinophils. Heterophilic granules are generally smaller than those of eosinophils and may not occupy all of the cytoplasm.
Heterophils range from 10 to 15 µm in diameter.

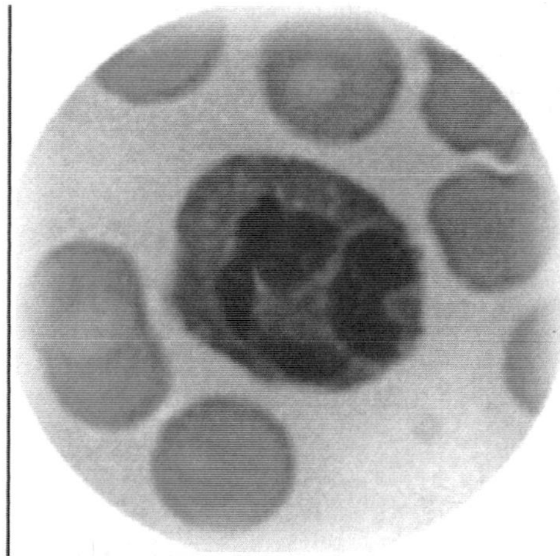

Eosinophil

Eosinophils are slightly larger than heterophils and are 12 to 16 µm in diameter.
The nucleus stains purple and often appears bilobed. Intensely acidophilic, round, cytoplasmic granules are present that are larger and more numerous than the granules in heterophils.

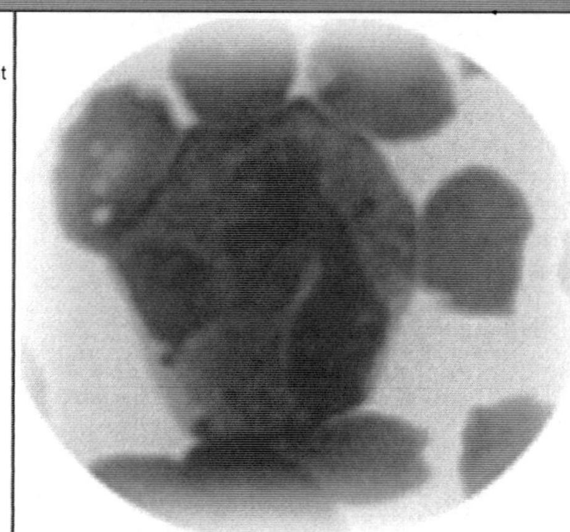

Monocyte

Monocytes are the largest circulating leukocytes in health and measure 15 to 18 µm in diameter. Monocytes have a large, variably shaped nucleus with chromatin that appears less condensed than that of heterophils. The cytoplasm is abundant and stains gray to blue-gray.
A few cytoplasmic vacuoles may be observed. Large, dark red granules have been described in the cytoplasm of some monocytes in association with nonspecific toxicity.

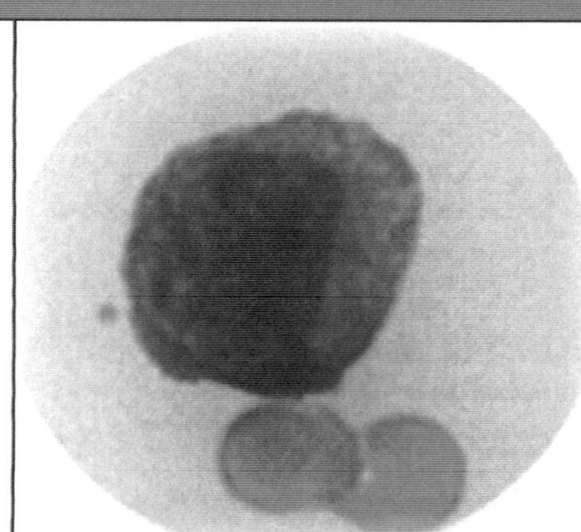

Basophil

Basophils have a light purple, lobulated nucleus and dark purple to purple-black cytoplasmic granules. They are approximately the same size as heterophils.	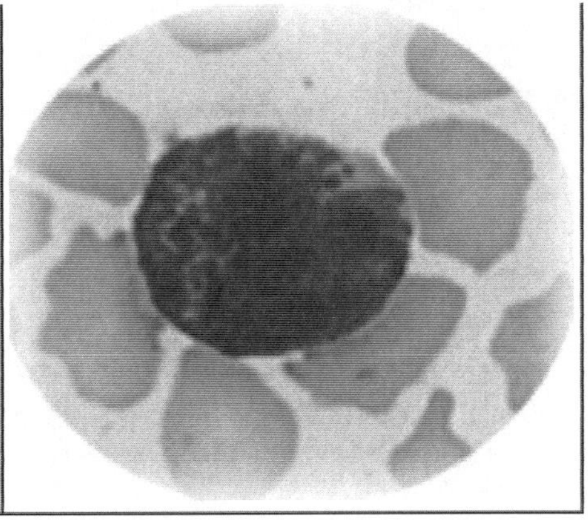

Images courtesy of UGA.

References

1. Negrete A, Lartigue C. Learning from education to communicate science as a good story. *Endeavour.* 2004;28:120–124.
2. Suckow MA, Schroeder V. *The Laboratory Rabbit.* 2nd ed. Boca Raton, FL: CRC Press/Taylor & Francis; 2010.
3. Marshall KL. Rabbit hematology. *Vet Clin North Am Exot Anim Pract.* 2008;11:551–567, vii.
4. Hinton M, Jones DR, Festing MF. Haematological findings in healthy and diseased rabbits, a multivariate analysis. *Lab Anim.* 1982;16:123–129.
5. Varga M. Chapter 2: Clinical pathology. In: Varga M. *Textbook of Rabbit Medicine.* 2nd ed. New York, NY: Butterworth Heinemann Elsevier; 2014:111–136.
6. Barthold SW, Griffey SM, Percy DH. *Pathology of Laboratory Rodents and Rabbits.* 4th ed. Ames, IA: Wiley-Blackwell; 2016:255.
7. Turner PV, Brash ML, Smith DA. Chapter 1: Rabbits. In: Turner PV, Brash ML, Smith DA. *Pathology of Small Mammal Pets.* Ames, IA. Wiley-Blackwell; 2018: 1–68.

SPEAKER INFORMATION
(click the speaker's name to view other papers and abstracts submitted by this speaker)

Joerg Mayer, DVM, MS, DABVP (Exotic Companion Mammal Practice), DECZM (Small Mammal), DACZM
Department of Small Animal Medicine and Surgery
College of Veterinary Medicine
University of Georgia
Athens, GA, USA

URL: https://www.vin.com/doc/?id=9768918

Speakers

B

Go to Table of Contents: Main: Speaker List

Andrew Bean, DVM, MPH, DABVP (Exotic Companion Mammal Practice)
Antimicrobial Resistance and Therapy in Exotic Pet Practice

Hugues Beaufrère, DVM, PhD, DACZM, DABVP (Avian), DECZM (Avian)
Clinical Lipidology in Psittacine Birds

Ethan Biswell, DVM, DACVP
The Reptilian and Amphibian Circulatory Systems: The Pathology

Heather A. Bjornebo, DVM, DABVP (Reptile & Amphibian Practice), CertAqV
Reptile Emergencies

João Brandão, LMV, MS, DECZM (Avian)
Clinical Endocrinology of Small Mammals

D

Go to Table of Contents: Main: Speaker List

Nicola Di Girolamo, DMV, MSc(EBHC), GPCert(ExAP), PhD, DECZM (Herpetology)
Practical Approach to Rabbit Anesthesia

Stephen J. Divers, BVetMed, DZooMed, DECZM (Herp), DECZM (ZHM), DACZM, FRCVS
The Reptile Liver: Clinical Investigation and Diagnosis

Grayson A. Doss, DVM, DACZM
Sedation in Birds
African Pygmy Hedgehogs: Husbandry and Clinical Management

E

Go to Table of Contents: Main: Speaker List

M. Scott Echols, DVM, DABVP (Avian)
The Avian Renal System from Form and Function to Clinical Perspectives

ExoticsCon
Program
Sponsors
Webinar Help

G

Go to Table of Contents: Main: Speaker List

Bonnie Gatson, DVM, DACVAA
Reptilian Cardiopulmonary Physiology and Inhalant Anesthesia: What Do We Know So Far?

H

Go to Table of Contents: Main: Speaker List

J. Jill Heatley, DVM, MS, DABVP, DACZM
Parrot Bornavirus: Current Clinical Summary of Diagnosis, Treatment, and Control
Diagnosis and Treatment of the Cardiovascular and Lymphatic Systems of Reptiles and Amphibians

Laurie Hess, DVM, DABVP (Avian Practice)
Companion Parrot Nutrition—Nutrient Requirements, Common Feeding Practices, and Nutrition-Related Diseases

J

Go to Table of Contents: Main: Speaker List

Dan H. Johnson, DVM, DABVP (Exotic Companion Mammal Practice)

M

Go to Table of Contents: Main: Speaker List

Christoph Mans, DrMedVet, DACZM, DECZM (ZHM)

Joerg Mayer, DVM, MS, DABVP (Exotic Companion Mammal Practice), DECZM (Small Mammal), DACZM

Jeffrey M.B. Musser, DVM, PhD, DABVP (Dairy)

O

Go to Table of Contents: Main: Speaker List

Susan Orosz, DVM, PhD, DABVP (Avian), DECZM (Avian)

Seth C. Oster, DVM

R

Go to Table of Contents: Main: Speaker List

Ellen Rasidi, BBiomedSc, BSc(Vet)(Hons), BVSc

Drury Reavill, DVM, DABVP (Avian and Reptile/Amphibian Practice), DACVP

Jenna Richardson

Anna-Marie Ruoff, DVM

S

Go to Table of Contents: Main: Speaker List

Stefania Scarabelli, DMV, DECVAA, DACVAA

Renée Schott, DVM, CWR

W

Go to Table of Contents: Main: Speaker List

Jeanette Wyneken, PhD

Association of Reptilian & Amphibian Veterinarians
5003 SW 41st Blvd
Gainesville FL 32608-4930
USA

ISBN 978-1-7138-3038-2

CURRAN ASSOCIATES INC.
proceedings.com